Springer Theses

Recognizing Outstanding Ph.D. Research

For further volumes:
http://www.springer.com/series/8790

Aims and Scope

The series "Springer Theses" brings together a selection of the very best Ph.D. theses from around the world and across the physical sciences. Nominated and endorsed by two recognized specialists, each published volume has been selected for its scientific excellence and the high impact of its contents for the pertinent field of research. For greater accessibility to non-specialists, the published versions include an extended introduction, as well as a foreword by the student's supervisor explaining the special relevance of the work for the field. As a whole, the series will provide a valuable resource both for newcomers to the research fields described, and for other scientists seeking detailed background information on special questions. Finally, it provides an accredited documentation of the valuable contributions made by today's younger generation of scientists.

Theses are accepted into the series by invited nomination only and must fulfill all of the following criteria

- They must be written in good English.
- The topic should fall within the confines of Chemistry, Physics, Earth Sciences, Engineering and related interdisciplinary fields such as Materials, Nanoscience, Chemical Engineering, Complex Systems and Biophysics.
- The work reported in the thesis must represent a significant scientific advance.
- If the thesis includes previously published material, permission to reproduce this must be gained from the respective copyright holder.
- They must have been examined and passed during the 12 months prior to nomination.
- Each thesis should include a foreword by the supervisor outlining the significance of its content.
- The theses should have a clearly defined structure including an introduction accessible to scientists not expert in that particular field.

Nicholas Andrew Wasley

Nano-photonics in III–V Semiconductors for Integrated Quantum Optical Circuits

Doctoral Thesis accepted by
the University of Sheffield, UK

Springer

Author
Dr. Nicholas Andrew Wasley
Department of Physics and Astronomy
University of Sheffield
Sheffield
UK

Supervisor
Prof. Maurice Skolnick
Department of Physics and Astronomy
University of Sheffield
Sheffield
UK

ISSN 2190-5053 ISSN 2190-5061 (electronic)
ISBN 978-3-319-01513-2 ISBN 978-3-319-01514-9 (eBook)
DOI 10.1007/978-3-319-01514-9
Springer Cham Heidelberg New York Dordrecht London

Library of Congress Control Number: 2013944518

Printed on acid-free paper

Springer is part of Springer Science+Business Media (www.springer.com)

Supervisor's Foreword

This thesis breaks new ground in the physics of photonic circuits for quantum optical applications. The photonic circuits are based either on ridge waveguides or photonic crystals with embedded quantum dots providing the single qubit, quantum optical emitters. The highlight of the thesis is the first demonstration of a spin-photon interface using an all waveguide geometry, a vital component of a quantum optical circuit, based on deterministic single photon emission from a single quantum dot. This work has recently been published in Physical Review Letters (January 2013), and is contained in Chaps. 4 and 5 of the thesis. Chapter 3 also contains an important contribution to the field, demonstrating the effects and limitations that inevitable disorder places on photon propagation in photonic crystal waveguides, a further key component of a quantum optical circuit. Overall, the thesis contains highly novel contributions to the field; such on-chip circuits are likely to be the only means to scale up the highly promising quantum dot-based quantum information technology. The quality of the thesis is a testament to Mr. Wasley's high experimental ability and also deep understanding of the relevant physics.

Sheffield, January 2013

Prof. Maurice Skolnick

Acknowledgments

I have been privileged to study in the Low Dimensional Structures and Devices (LDSD) group of the Department of Physics and Astronomy at the University of Sheffield, alongside many talented individuals.

I would like to thank those members of the LDSD group with whom I have shared my time during my studies, including past and present postgraduate students as well as the academic and support staff.

In particular, I would like to thank my supervisor Maurice Skolnick for all his efforts as group leader and providing guidance and support throughout my time in Sheffield.

I must also thank Isaac Luxmoore for his hard work, patience and the exemplary manner in which he conducts his research, from whom I have learned a lot and to whom I owe much.

Finally, I would like to thank my family and friends for their continued support, especially Laura who has had to suffer me the most during my studies.

Abstract

This thesis describes the optical spectroscopic measurements of III-V semiconductors used to investigate a number of issues related to the development of integrated quantum optical circuits.

The disorder-limited propagation of photons in photonic crystal waveguides in the slow light regime is investigated. The analysis of Fabry–Perot resonances is used to map the mode dispersion and extract the photon localisation length. Anderson-localised modes are observed at high group indices, when the localisation lengths are shorter than the waveguide lengths, consistent with the Fabry–Perot analysis.

A spin-photon interface based on two orthogonal waveguides is introduced, where the polarisation emitted by a quantum dot is mapped to a path-encoded photon. Operation is demonstrated by deducing the spin using the interference of in-plane photons. A second device directly maps right and left circular polarisations to anti-parallel waveguides, surprising for a non-chiral structure but consistent with an off-centre dot.

Two dimensional photonic crystal cavities in GaInP and full control over the spontaneous emission rate of InP quantum dots is demonstrated by spectrally tuning the exciton emission energy into resonance with the fundamental cavity mode. Fourier transform spectroscopy is used to investigate the short coherence times of InP quantum dots in GaInP photonic crystal cavities.

Additional technological developments are also presented including a quantum dot registration technique, electrical tuning of quantum dot emission and uniaxial strain tuning of H1 cavity modes.

Contents

Publications

I. J. Luxmoore, E. Daghigh Ahmadi, **N. A. Wasley**, A. M. Fox, A. I. Tartakovskii, A. B. Krysa and M. S. Skolnick. "Control of spontaneous emission from InP single quantum dots in GaInP photonic crystal nanocavities". *Applied Physics Letters* **97**, 181104 (2010).

I. J. Luxmoore, E. Daghigh Ahmadi, B. J. Luxmoore, **N. A. Wasley**, A. I. Tartakovskii, M. Hugues, M. S. Skolnick and A. M. Fox. "Restoring mode degeneracy in H1 photonic crystal cavities by uniaxial strain tuning". *Applied Physics Letters* **100**, 121116 (2012).

N. A. Wasley, I. J. Luxmoore, R. J. Coles, E. Clarke, A. M. Fox and M. S. Skolnick. "Disorder-limited photon propagation and Anderson-localization in photonic crystal waveguides". *Applied Physics Letters* **101**, 051116 (2012).

I. J. Luxmoore, **N. A. Wasley**, A. J. Ramsay, A. C. T. Thijssen, R. Oulton, M. Hugues, S. Kasture, Achanta V. G, A. M. Fox and M. S. Skolnick. "Interfacing spin in an InGaAs quantum dot to a semiconductor waveguide circuit using emitted photons". *Physical Review Letters* **110**, 037402 (2013).

I. J. Luxmoore, R. Toro, O. Del Pozo-Zamudio, **N. A. Wasley**, E. A. Chekhovich, A. M. Sanchez, R. Beanland, A. M. Fox, M. S. Skolnick, H. Y. Liu and A. I. Tartakovskii. "III-V quantum light source and cavity-QED on Silicon". *Scientific Reports* **3**, 01239 (2013).

Chapter 1
Introduction

1.1 Integrated Quantum Optical Circuits and Networks for Quantum Information Processing

Proposals for a practical implementation of quantum information processing (QIP) are many and varied. This reflects both the broad interest and the numerous challenges that remain in the development of functional systems.

A system based on III–V semiconductor materials has been an early front runner in the development of a quantum information processor due to the electrical and optical properties provided by III–V systems and the mature fabrication technologies that have been developed by the III–V based opto-electronics industry.

Progress toward the control and manipulation of III–V quantum dots (QDs) for use as static qubits has been good, and there is now a strong demand to develop the tools required to link remote qubit nodes. The integration of static nodes (QDs) and flying qubits (photons) to process and transfer quantum information in scalable III–V systems spans research fields in nano-photonics and quantum optics.

This research toward scalable QIP using III–V semiconductors is currently driven by the construction of simple quantum-optical circuits capable of demonstrating a number of principles that will form the building blocks of future QIP networks, and thus demonstrate the first fully functional integrated circuits.

This thesis focuses on a number of areas in nano-photonics and quantum-optics that are relevant to the development of integrated quantum optical circuits from III–V semiconductors.

It begins with a chapter that contains an introduction to the general principles of quantum computing and the associated physics for implementing quantum information processing using III–V semiconductors including self assembled quantum dots and photonic crystals.

Following this introductory chapter, Chap. 2 contains the details of the samples, spectroscopic techniques and apparatus used to perform the measurements presented in the subsequent chapters.

N. A. Wasley, *Nano-photonics in III–V Semiconductors for Integrated Quantum Optical Circuits*, Springer Theses, DOI: 10.1007/978-3-319-01514-9_1,

Chapter 3 investigates the propagation length of photons in photonic crystal waveguides propagating in the slow-light regime by analysing Fabry-Perot resonances and links this to the observation of Anderson-localised modes.

Chapter 4 introduces a spin-photon interface based on two orthogonal waveguides, where the polarisation emitted by a quantum dot is mapped to a path-encoded photon, for the purpose of communicating the spin state between two points on a chip. Operation is demonstrated by deducing the spin using the interference of in-plane photons. A second device presented in Chap. 5 directly maps right and left circular polarisations to anti-parallel waveguides, surprising for a nonchiral structure but consistent with an off-center dot, as supported by computer simulations of the structure.

Chapter 6 presents experiments conducted in GaInP photonic crystal cavities where full control over the spontaneous emission rate of InP quantum dots is demonstrated by spectrally tuning the exciton emission energy into resonance with the fundamental cavity mode. Fourier transform spectroscopy is used to investigate the short coherence times of InP quantum dots in GaInP photonic crystal cavities.

Chapter 7 presents a number of additional technological approaches developed to assist in the fabrication and design of integrated circuits including quantum dot registration, electrical tuning of QD emission and uniaxial strain tuning of H1 cavity modes to restore the fundamental mode degeneracy.

Finally, Chap. 8 presents a summary of the conclusions drawn from the experimental results and discusses possible future directions in which work in this area could be taken.

1.1.1 Quantum Information Processing

In a classical computer information is encoded in 'bits', where the 'bit' can take one of two possible values, 0 or 1. Information is processed by performing a series of logical operations that return the values 0 or 1. It has been proposed that information processing can take place using fundamentally different principles based on the effects of quantum mechanics. In such a system the 'bit' is replaced by a 'quantum bit' or 'qubit' [1].

A qubit consists of the orthogonal states $|0\rangle$ and $|1\rangle$ and can, in principle, be constructed from any quantum two-level system. Unlike the bit, the qubit can exist in any linear superposition of the two states described by a wavefunction $|\psi\rangle = \alpha|0\rangle + \beta|1\rangle$, where the α and β are complex amplitudes with a normalisation condition that $|\alpha|^2 + |\beta|^2 = 1$.

Information can then be processed by manipulating single or multiple qubits using quantum operations or quantum logic gates. Some of the quantum logic gates, such as the Hadamard gate that is responsible for converting a pure $|0\rangle$ or $|1\rangle$ state into a superposition state, have no classical analogue.

The manipulation of a superposition state using qubit gates allows the output for the states $|0\rangle$ and $|1\rangle$ to be calculated in parallel allowing a quantum information processor to calculate the result for many inputs simultaneously.

This, and other, quantum mechanical effects such as entanglement, not possible in classical systems have been shown to be useful for efficiently solving certain types of problems including finding the prime factors of large numbers. Shor's algorithm is a well known example of a quantum information process that is designed to find the prime factors of large numbers [2]. It has been shown that Shor's algorithm is capable of finding the prime factors of large numbers exponentially faster than the most efficient known classical factoring algorithm.

1.1.2 Implementation of QIP in the Solid State

The concept of quantum information processing requires only that the qubit be a quantum two-level system. However, for the practical realisation of quantum computing a number of more general requirements are summarised by the DiVincenzo criteria [3].

An example of a requirement stipulated by the DiVincenzo criteria is that the qubit must have long coherence times relative to the gate operation time. On the basis of these criteria QIP schemes have been proposed in a number of different physical systems that have the potential to satisfy these requirements.

The first proposals for optical quantum information processing devices were based on cavity quantum electrodynamics (CQED) in atomic systems [4, 5]. These systems consist of macroscopic optical resonators coupled to atoms or ions that are interconnected via optical beams propagating through free space. However, their high complexity and macroscopic size prohibit their scalability.

Another solution is to implement systems that use solid state photonics and III–V semiconductors, which provide the stability of the solid state material system and the potential for scalability.

Although several options exist for suitable optical two-level emitters in the solid state like nitrogen vacancy (NV) centers in diamond [6], III–V quantum dots are a promising candidate as they interact strongly with light due to a large optical dipole, have sharp atomic like emission lines that can be tuned using band structure engineering and they can be scaled up and integrated with photonic structures to form on-chip devices using mature fabrication technologies developed for integrated III–V based circuits.

A number of the required components for solid state QIP already exist. These include optical nanocavities (microdisks [7, 8], photonic crystals [9] and micropillar cavities [10]), efficient production and transport of single photons [11, 12] and the optical or electrical control and manipulation of QD based qubits [13, 14].

1.1.3 Quantum Optical Circuits

Fundamental to the idea of quantum information processors is the ability to convert quantum states from one physical system to those of a another in a reversible manner to transfer quantum information through space. Such quantum connectivity in networks can be achieved by the optical interactions of single photons and quantum dots to distribute entanglement across the network and the teleportation of quantum states between nodes.

In proposed III–V systems, quantum information is generated, processed and stored locally in quantum dots (nodes). These nodes are then linked by quantum channels which transport quantum states from site to site with high fidelity and distribute entanglement across the entire network. The transfer of quantum information using photons as flying qubits in a scalable geometry requires optical circuitry to control, confine and manipulate the photons.

A number of technologies exists that can achieve this. At its simplest, nanowires of semiconductor material can control the photon path using total internal reflection, similar in function to an optical fiber.

Photonic crystals are a more sophisticated technology that operate using a different principle. In comparison to semiconductor nanowires, they offer increased mechanical stability and a different functionality that provides a route to cavity quantum electrodynamics (CQED) [9].

1.2 Photonic Crystals

Photonic crystals are a two dimensional periodic arrangement of materials with differing dielectric constants. The periodicity of the structures cause Bragg reflections that control the propagation properties of light through the crystal. Depending on the dielectric constant, length of periodicity and type of symmetry the photonic crystal possesses, the propagation of light with certain wavelengths can be prohibited due to the destructive interference caused by the Bragg reflections [15]. The reflections form a bandgap in the dispersion relation for in plane light propagation, known as a photonic bandgap, analogous to the electronic bandgap in semiconductors.

Photonic crystals in one, two and three dimensions have been fabricated in a number of different material systems [16]. The types of photonic crystal employed in our work are two dimensional photonic crystal slabs, the interest in which originates from their potential for integration in planar quantum optical circuits which provides an inherently scalable geometry that can be fabricated using mature technologies.

The two dimensional photonic crystal slabs used in the experiments detailed in this thesis take the form of a periodic triangular lattice of low index air holes, with lattice constant, a, and hole radius, r, patterned through a suspended membrane of high index semiconductor material. A scanning electron microscope (SEM) image of this type of photonic crystal is presented in Fig. 1.1a.

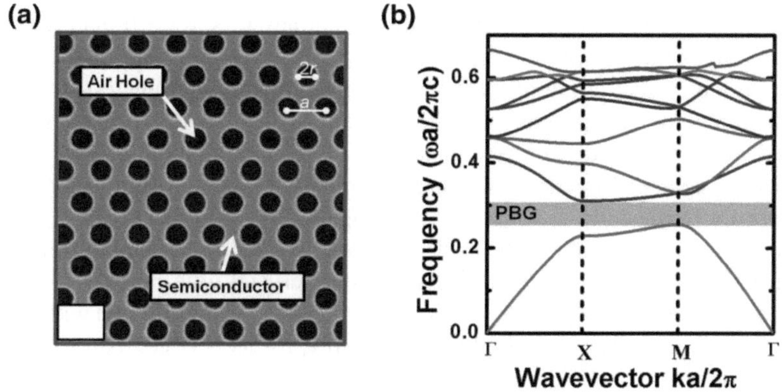

Fig. 1.1 **a** Scanning electron microscope (SEM) image of a photonic crystal slab where air holes of radius, r, have been etched through semiconductor in a periodic triangular lattice with lattice constant, a. **b** A band diagram calculated using the MIT photonic bands software package (MPB) for a photonic crystal with $r/a = 0.28$. A photonic bandgap, highlighted in *grey*, forms between band 1 (*red*) and band 2 (*blue*)

1.2.1 Photonic Bandstructure of a Photonic Crystal

The angular frequency of an electromagnetic mode, ω, is related to the wavevector, k by $\omega = ck = c\sqrt{k_x^2 + k_y^2 + k_z^2}$, where k_x, k_y and k_z are orthogonal components of k and c is the speed of light. In free space, electromagnetic modes approximate plane waves where the spectrum of available states is continuous for all frequencies above the light line $\omega = ck_x$, which is called the light cone. For any given value of k_x there will exist modes with every possible frequency greater than ck_x because k_y and k_z may take any value.

For a homogeneous planar structure that has a refractive index larger than the free space value, new electromagnetic solutions will exist within the structure below the light line. As the refractive index is larger in the structure than in free space these modes have frequencies that are lower relative to the corresponding modes in free space.

We define $k_\perp$ and $k_\parallel$ as the wavevectors perpendicular and parallel to the interface between the high index planar structure and free space. The only free space solutions for modes below the light line $w = ck_\parallel$, are those with imaginary $k_\perp = \pm i\sqrt{k_\parallel^2 - \omega^2/c^2}$. These are evanescent electric fields that decay exponentially away from the material. These states localised in the vicinity of the high refractive index material are index guided modes.

Because these modes are localised in one direction they form a discrete set of frequencies for a given value of $k_\parallel$ and we obtain discrete bands of $\omega(k_\parallel)$ below the light line. The discrete bands of conserved values for frequency and wavevector define the photonic bandstructure of the system. This band structure in the specific case of a homogeneous material in the absence of any loss, reflection or diffraction is referred to as the free photon model.

The periodic modulation of dielectric permittivity in a photonic crystal causes a perturbation to this free photon model that separates the dispersion curves which can lead to a photonic bandgap in the bandstructure of the photonic crystal. This is a range of frequencies for which there are no supported modes within the crystal, and therefore are frequencies at which light propagation is forbidden [17].

The MIT photonic bands (MPB) [18] software package was used to calculate the bandstructure for TE modes in a GaAs photonic crystal with $r/a = 0.28$ and the results are presented in Fig. 1.1b. The individual bands are highlighted with different colours. A photonic bandgap, highlighted in grey, forms between band 1 (red) and band 2 (blue).

1.2.2 Photonic Crystal Nano-cavities

If a defect in the crystal lattice is introduced to the photonic crystal by altering or removing one or more of the air holes, it can create an optical cavity within the crystal.

In a one dimensional approximation, an optical cavity in a material with refractive index n, has resonances when the cavity length L_{cav} is equal to an integer multiple, m, of the intra-cavity half wavelength, λ,

$$L_{cav} = \frac{m\lambda}{2n} \tag{1.1}$$

This resonance occurs when there are in-phase reflections of light around the cavity. The constructive interference of wavelengths at the resonance conditions produce modes of high field amplitude. Wavelengths of light that satisfy these resonance conditions are known as cavity modes.

By designing a defect within the photonic crystal to support cavity modes with wavelengths that lie within the photonic band gap of the crystal we are able to confine light within the plane of the crystal [19]. Total internal reflection at the semiconductor air interfaces provides further confinement in the direction perpendicular to the plane of the crystal.

The design of the photonic crystal cavity determines the properties of the modes that it supports, such as wavelength, polarisation and light confinement [20, 21]. The ability of a cavity to confine light is an important property when considering the applications in QIP.

The confinement ability or quality factor (Q-factor) of a cavity is defined by two parameters that are important concepts for the analysis of QD-cavity interactions that follow. They are the frequency of the resonant mode, ω, and the lifetime of a photon in the cavity, τ. The lifetime is how long a photon within the cavity can exist before escaping. In a perfect optical cavity photons never escape and have an infinite

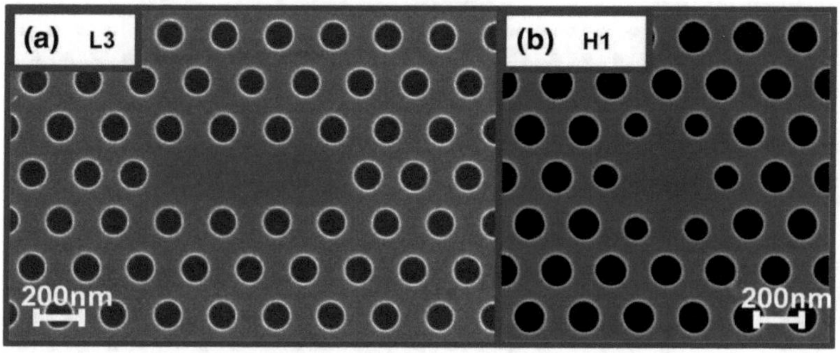

Fig. 1.2 SEM images of (**a**) L3 and (**b**) H1 type photonic crystal cavities. With the etched air holes and the cavity formed by the omission of three and one air holes respectively

lifetime. In practice a photon will have a finite lifetime within the cavity. A cavity decay rate is defined $\kappa = \frac{1}{\tau_{cavity}}$, such that the Q-factor of a given cavity is defined $Q = \frac{\omega}{\kappa}$.

Most photonic crystal cavity designs are based on the removal of one or more holes in periodic lattice of air holes. Two of the most well studied of these types of cavities over the past ten years have been the H1 and L3 cavities, which remove one and three holes from the lattice respectively [22, 23]. Both cavities can support a number of modes within the photonic band gap. Scanning electron microscope (SEM) images of an L3 and H1 type cavity are shown in Fig. 1.2.

1.2.3 Guided Modes in Photonic Crystal Waveguides

A point defect within a photonic crystal is a localised disruption to the periodicity of the dielectric modulation that can act as an optical resonator capable of supporting electromagnetic modes of discrete frequencies within the photonic bandgap. A photonic crystal waveguide can be considered as regularly spaced point defects within the photonic crystal, where coupling between the regularly spaced point defects leads to the formation of a transmission band or guided mode where k is still conserved within the photonic bandgap of the crystal. Hence light may propagate along the defect confined by total internal reflection in the vertical direction and Bragg reflection in the lateral direction.

Removing a single row of holes from a photonic crystal produces a single mode waveguide, such that there is at most one guided mode for a given frequency within the bandgap. This is important for quantum information processing applications where a signal may become scrambled if more than one mode of propagation exists. This can be the case in multimode waveguides produced by removing more than one row of holes or by modifying the position of the holes next to the waveguide see 3.10.2. An SEM image of a W1 type photonic crystal waveguide is displayed in Fig. 1.3.

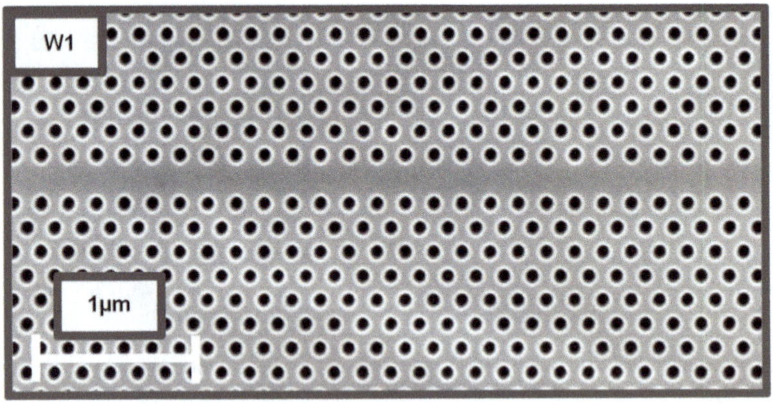

Fig. 1.3 SEM image of an W1 type photonic crystal waveguide. With the etched air holes and the waveguide formed by the omission of a single row of air holes

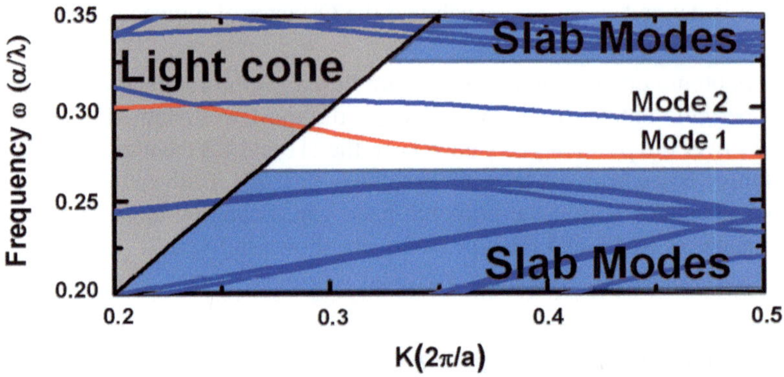

Fig. 1.4 A band diagram calculated using MPB for a W1 GaAs PCW with r/a = 0.28. The mode 1 (*red*) indicates the first guided mode studied in this chapter. The light line w = ck and light cone are shaded *grey* whilst other modes not localised to the PCW or outside the bandgap are labelled slab modes and shaded *blue*

The guided waveguide modes for a given structure can be calculated by computing definite-frequency eigenstates of Maxwell's equations in periodic dielectric structures. The MIT photonic bands (MPB) [18] software package was used to calculate the bandstructure for the PCWs and the results are displayed in Fig. 1.4.

The gap guided mode highlighted in red (Mode 1) is the subject of our investigation in Chap. 3. It is composed of contributions from two types of guided mode that anti-cross. The anti-crossing is a result of coupling between a refractive index guided mode and a mode guided by the PBG [24, 25]. The mode produces slow light in the region toward the edge of the Brillouin zone.

Slow-light

Slow light is a term used to describe light that travels through a medium at a group velocity many times lower than the speed of light in a vacuum, c. The group velocity of light is given by the inverse of the first order dispersion. It is important for QIP applications because it offers a route by which light-matter interactions may be enhanced. However, it also has potential applications in optical buffering, phase shifting [26] and temporary storage processes which at present rely on solutions that have a slow response times such as mechanical variable delay lines or delay lines that incorporate an optical switch [27]. Slow light was first experimentally observed in silicon based W1 photonic crystal waveguides by Notomi in 2001 [24].

The periodic modulation of the dielectric permittivity at the scale of optical wavelengths that is responsible for forming the bound waveguide modes also results in a density of states that differs from that of free space. The photonic density of states is defined as the number of electromagnetic modes available per unit frequency. In an ideal isotropic crystal, maxima in the density of states occur at the photonic band edge. Secondary maxima occur each time the dispersion curves of guided modes within the photonic bandgap present a tangent parallel to the axis of the wave vector ($\frac{d\omega}{dk} = 0$) [15]. Hence the density of photonic states is high where the group velocity of light in the guided mode falls to zero. As such the slow-light region of the guided mode highlighted in Fig. 1.4 occurs toward the edge of the Brillouin zone where $d\omega/dk$ falls to zero and the mode looks 'flat'. This region can be used to provide a Purcell enhancement to the radiative emission of QDs.

1.3 Quantum Dots

Quantum dots are islands of low bandgap semiconductor material typically 20 nm wide and 5 nm in height that are surrounded by a semiconductor material of higher bandgap. This creates an effective potential well that can trap electrons and holes.

The spatial confinement of the quantum dot forces the valence and conduction bands into discrete energy levels. This quantum confinement provides the QD with atom-like properties; as such it behaves like a two-level emitter.

Quantum dots can be designed to operate at various wavelengths spanning the entire near infrared spectrum by engineering the bandgap of the QD. This can be achieved by controlling the size of the quantum dot during growth and more commonly the material from which the QDs are made.

For applications in quantum information science it is desirable to operate below the wavelength of 1 μm because of the high performance silicon photodetectors available in this wavelength range. However, in QDs operating at short wavelengths, the ground state energy of the quantum dot is very close to the GaAs bandgap, so carriers confined in the QD can be scattered out by phonon interactions at energy levels of kT (k the Boltzmann constant and T the temperature). To minimize the effect of phonon scattering, the QDs must be cooled down to temperatures below 50 K.

The quantum dots used in this thesis were made from two distinct material systems. Indium arsenide QDs grown on host gallium arsenide material (InAs/GaAs) and indium phosphide quantum dots grown in gallium indium phosphide (InP/GaInP). InAs has a bandgap energy of 0.36 eV which produces quantum dots with emission at a wavelength of around 970 nm, and is a well studied system in photonic crystals. Indium phosphide quantum dots grown in gallium indium phosphide emit at shorter wavelengths and are relatively little studied in photonic crystal applications. An InP bandgap energy of 1.35 eV produces QDs with emission at a wavelengths of around 670 nm. This not only puts their emission into the visible range but also at a wavelength that improves the efficiency of commercially available silicon detectors compared with InAs QD emission. The reduced wavelength of emission however requires a reduction in the size of photonic crystal cavity features. Fabrication errors are not proportionally reduced so there tends to be a reduction in the Q values achievable in the cavities.

1.3.1 Energy States

A QD is a finite three dimensional potential well. The spatial confinement forces trapped electrons and holes into discrete energy levels. Trapped electrons and holes form bound excitonic states. It is because of this behaviour that quantum dots are sometimes referred to as 'artificial atoms', and provides the discrete two level system that makes bound states in QDs suitable for a number of cavity QED and QIP applications.

Conduction band electrons carry spin component $S = \pm\frac{1}{2}$ and heavy holes $S = \pm\frac{3}{2}$ (strain within quantum dots caused by lattice mismatch modifies the electronic band structure. This lifts the degeneracy of light hole and heavy hole states in the valence band, with a typical splitting of $\sim$30 meV. The light hole states are therefore neglected when discussing the lower energy states of quantum dots). This gives a total angular momentum projection for lowest energy electron-hole pairs of $M = \pm1, \pm2$. The states with $M = \pm2$ are known as dark states and cannot recombine optically since they require an angular momentum transfer of $2\hbar$. The $M = \pm1$ state gives rise to the neutral exciton X^0 transition.

The isotropic electron-hole exchange interaction couples the spins of the electron and hole, that results in a splitting of the dark and bright states. The anisotropic exchange interaction which is caused by asymmetry of the QD lifts the degeneracy of both dark and bright manifolds, producing four excitonic eigenstates, known as the fine structure. $|D_{a,b}\rangle = (|+2\rangle \pm |-2\rangle)/\sqrt{2}$, the dark doublet and $|X^0_{x,y}\rangle = (|+1\rangle \pm |-1\rangle)/\sqrt{2}$, the bright doublet.

The fine structure splitting of the bright exciton doublet is greater than that of the dark exciton doublet due to the influence of the long range part of the exchange

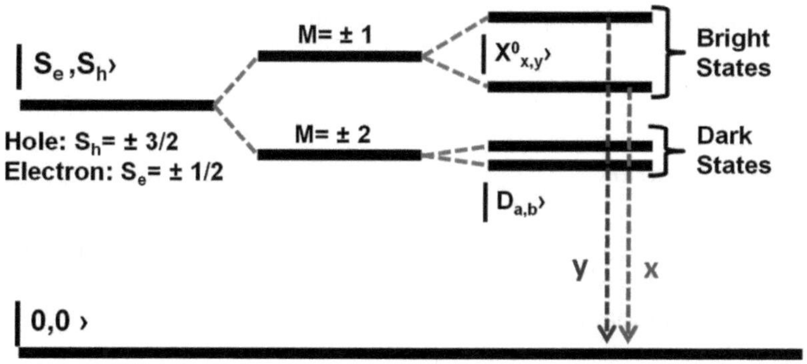

Fig. 1.5 Energy level scheme of the eigenstate splitting for a neutral exciton. With spins, S, and total angular momentum projection for the states, M, bright and dark manifolds $|X^0_{x,y}\rangle$, $|D_{a,b}\rangle$ and the orthogonally linearly polarised emission Y, X are labelled

interaction in an asymmetric QD, which influences only the bright states. This acts to enhance the splitting provided by the short range part of the exchange interaction which influences both the bright and dark states in asymmetric QDs, as detailed in Ref. [28].

The two states of the bright doublet decay by emitting photons that are orthogonally polarised to one another, a schematic representation of the excitonic energy states in a QD is presented in Fig. 1.5.

The polarisation of the excitonic states can have a large influence on the observed properties of the system and the way in which it can be manipulated. For instance the fundamental cavity mode of an L3 type cavity is linearly polarised along the direction perpendicular to the defect line. This gives maximum coupling between the cavity mode and corresponding polarised bright exciton state. The second bright state is left uncoupled to the cavity.

Magnetic Fields

The application of an external magnetic field can be used to further lift the degeneracy of quantum dot spin states. In a magnetic field the fine structure splitting is increased by the Zeeman interaction between the electron and hole spins with the magnetic field, B. The magnetic field causes a Zeeman splitting between the spin states, $\Delta E_Z = g\mu_B B$, where μ_B is the Bohr magneton and g is the exciton g-factor.

At low and zero magnetic fields the fine structure eigenstates are linear combinations of the circularly polarised exciton. At high fields the Zeeman interaction is much larger than the anisotropic exchange interaction and the linearly polarised states transform into circularly polarised states. This transition reflects the restoration of the QDs rotational symmetry by the magnetic field. A full description of this transition is presented in Ref. [28]. This description applies to experiments in which the magnetic field is applied along the growth direction, known as the Faraday geometry.

1.3.2 QDs as Qubits

An exciton confined in a QD is a potential qubit, and full quantum control has been demonstrated [29]. However, an exciton has a short coherence time that is limited by radiative recombination to about a nanosecond, due to the large overlap of the electron and hole wavefunction in the quantum dot.

There has also been interest in using carrier spins as qubits. However, it is more difficult to control a carrier spin with a fast optical pulse, because the spin does not interact directly with photons. Despite this challenge, electron spin lifetimes of up to a ms have been reported [30] and optical spin rotations of single electron spins have been recently demonstrated [13].

Although the intrinsic coherence time of an electron spin is long, the extrinsic dephasing time is much smaller, just a few nanoseconds. This has stimulated interest in the use of a hole spin in a QD as a potential qubit because the hole's p-type wave-function, leads to a reduced hyperfine interaction responsible for the decoherence process [31]. Full quantum control of a single hole spin confined in a self assembled quantum dot has also recently been demonstrated [32].

1.4 Cavity QED

The interaction between light confined within optical cavities and atomic, QD or other particle states is known as cavity quantum electrodynamics (QED). This interaction can be considered as an interaction between a two level quantum emitter and the electromagnetic field. In our systems this is the interaction between a quantum dot with a ground and excited state and the electric field due to a single photon with energy, hc/λ, in a cavity or waveguide.

When interacting with the light field, transitions between the ground and excited states of the emitter take place at a rate, W, described by Fermi's Golden Rule.

$$W = \frac{2\pi}{\hbar}|M_{12}|^2 g(\omega) \tag{1.2}$$

where M_{12} is the transition matrix element,

$$|M_{12}| = -\vec{\mu}_{12}.\vec{E}_{\vec{r}} \tag{1.3}$$

in which $\vec{\mu}_{12}$ is the electric dipole moment associated with the transition and is considered to be fixed in our analysis, $\vec{E}$ is the electric field strength that is a function of space, $\vec{r}$, and $g(\omega)$ is the density of available final states for transitions with frequency, ω.

The energy in the electric field due to one photon of energy, hc/λ, is fixed. If the electric field is confined to a cavity the electric field energy in the cavity mode in which the photon propagates is defined as

$$\int_{V_{mode}} \epsilon_{\vec{r}} |\vec{E}_{\vec{r}}|^2 dV = hc/\lambda = \varepsilon_{vac}^2 \tag{1.4}$$

where V_{mode} is the volume occupied by the mode, ε_{vac} is the magnitude of the vacuum field and $\epsilon_{\vec{r}}$ is the dielectric constant which is a function of position as the modal volume for a photonic crystal cavity incorporates transitions between the semiconductor and air.

Given the transition rate defined in Eq. 1.2 the probability of a single transition taking place can be increased in two ways. Firstly, by increasing the interaction time between the photon and the emitter. This is achieved by confining the light within the cavity for longer. This requires a reduction in the cavity decay rate κ, and an increase in Q.

Secondly, if the volume of the cavity is reduced, the volume of the cavity modes are reduced. As the energy in the electric field is fixed for a given photon energy the reduction in cavity volume causes an increase in the strength of the electric field confined in the cavity, $\vec{E}_{\vec{r}}$. It is for these reasons that photonic crystal cavities, that can support high Q-factors, (typically in the order of 8000–10000 for our GaAs based photonic crystals and up to a maximum of 7500 in our GaInP crystals) with small modal volumes are highly favourable systems for studying cavity QED.

1.4.1 Weak Coupling: Purcell Effect

In Fermi's Golden Rule the transition rate is proportional to the density of available final states.

$$W \propto g(\omega) \tag{1.5}$$

In a cavity the availability of states is significantly modified compared to free space due to the limited modes that are supported. If we assume there is only one mode supported within a cavity then

$$\int_0^\infty g(\omega) d(\omega) = 1 \tag{1.6}$$

given the frequency for the emitter, ω_0, with a half width $\Delta\omega_0$ a normalised Lorentzian fit satisfies this as the density of available states for the emitter

$$g(\omega_0) = \frac{2}{\pi \Delta\omega_c} \frac{\Delta\omega_c^2}{4(\omega_0 - \omega_c)^2 + \Delta\omega_c^2} \tag{1.7}$$

The number of available states is strongly affected by the difference between the frequency of the cavity mode and the frequency of the emitter, $\omega_0 - \omega_c$. We define the Purcell factor, F_p, as the ratio of transition rates for an emitter within a cavity, W_{cavity}, and in free space, W_{free}.

$$F_p = \frac{W_{cavity}}{W_{free}} = \frac{3Q(\lambda/n)^3}{4\pi^2 V_0} \xi^2 \frac{\Delta\omega_c^2}{4(\omega_0 - \omega_c)^2 + \Delta\omega_c^2} \tag{1.8}$$

where ξ is the normalised dipole orientation

$$\xi = \frac{|\vec{p}.\vec{E}|}{|\vec{p}||\vec{E}|} \tag{1.9}$$

$\vec{p}$ is the electric dipole of the QD, such that

$$|M_{12}|^2 = |\vec{\mu_{12}}.\vec{E}_{(\vec{r})}|^2 = \xi^2 \mu_{12}^2 \varepsilon_{vac}^2 \tag{1.10}$$

This modification of transition rates caused by an optical cavity is known as the Purcell effect. A Purcell factor greater than one indicates an enhancement in the relative rate of transition. A Purcell factor less than one indicates a suppression in the relative rate of transition.

For a resonant emitter and mode optimally located at the antinode of the cavity mode field, the Purcell factor reduces to

$$F_p = \frac{3Q(\lambda/n)^3}{4\pi^2 V_0} \tag{1.11}$$

The Purcell effect is the dominant process observed in the weak coupling regime of cavity QED.

1.4.2 Strong Coupling: Rabi Splitting

If we define a coupling parameter, g_0, that represents the interaction energy between the QD electric dipole moment $\vec{\mu}_{12}$ and the electric field $\vec{E}$

$$g_0 = \frac{1}{\hbar} |\vec{\mu}_{12}.\vec{E}| \tag{1.12}$$

The system is well described by the Jaynes-Cummings Hamiltonian, the solutions of which produce two energy Eigenstates separated by a factor dependent on the strength of the coupling

$$\Delta E = 2\sqrt{g_0^2 - \frac{(\kappa - \gamma)^2}{16}} \tag{1.13}$$

where κ is the photon decay rate of the cavity as previously defined and γ is the non-resonant decay rate of the emitter into modes other than the cavity mode [33].

When $g_0 \ll (\kappa, \gamma)$ this factor is imaginary and there is no energy splitting of the two states. This is the weak coupling regime and the Purcell effect is the dominant phenomenon. When $g_0 \gg (\kappa, \gamma)$, the photon is emitted and reabsorbed by the quantum dot before it is lost from the cavity. This factor becomes real and the degeneracy of the states is lifted forming two polariton states. An upper polariton state $|p_+\rangle$ and a lower polariton state $|p_-\rangle$. This is the strong coupling regime. The magnitude of the energy splitting, ΔE, is known as the vacuum Rabi Energy.

The signature of a transition between the weak coupling regime and strong coupling regime is the splitting of the QD peak and cavity mode peak when in resonance. Experimentally this is identified by an anti-crossing of the two emission peaks as they are tuned across one another [7, 9, 10].

References

1. M.A. Nielson, I.L. Chuang, *Quantum Computing and Quantum Information* (Cambridge Uniiversity Press, Cambridge, 2000)
2. P. Shor, Algorithms for quantum computation: discrete logarithms and factoring. in *Proceedings of the 35th Annual Symposium on Foundations of Computer Science*, (IEEE Computer Society Press, Los Alamitos, 1994), pp. 124–134
3. D.P. DiVincenzo, The Physical Implementation of Quantum Computation, arXiv, preprint quant-ph/0002077, Feb. 2000
4. J. Cirac, P. Zoller, H. Kimble, H. Mabuchi, Quantum state transfer and entanglement distribution among distant nodes in a quantum network. Phys. Rev. Lett. **78**, 3221–3224 (Apr. 1997)
5. L.M. Duan, H. Kimble, Scalable photonic quantum computation through cavity-assisted interactions. Phys. Rev. Lett. **92**, 1–4 (Mar. 2004)
6. C. Santori, P.E. Barclay, K.M.C. Fu, R.G. Beausoleil, S. Spillane, M. Fisch, Nanophotonics for quantum optics using nitrogen-vacancy centers in diamond. *Nanotechnology*, vol. 21, (2010), p. 274008
7. E. Peter, P. Senellart, D. Martrou, A. Lemaître, J. Hours, J. Gérard, J. Bloch, Exciton-photon strong-coupling regime for a single quantum dot embedded in a microcavity. Phys. Rev. Lett. **95**(6), 1–4 (2005)
8. B. Gayral, J.M. Gérard, A. Lemaître, C. Dupuis, L. Manin, J.L. Pelouard, High-Q wet-etched GaAs microdisks containing InAs quantum boxes. Appl. Phys. Lett. **75**(13), 1908 (1999)
9. T. Yoshie, A. Scherer, J. Hendrickson, G. Khitrova, H.M. Gibbs, G. Rupper, C. Ell, O.B. Shchekin, D.G. Deppe, Vacuum Rabi splitting with a single quantum dot in a photonic crystal nanocavity. Nature, **432**, 200–203 (2004)
10. J.P. Reithmaier, G. Sek, A. Löffler, C. Hofmann, S. Kuhn, S. Reitzenstein, L.V. Keldysh, V.D. Kulakovskii, T.L. Reinecke, A. Forchel, Strong coupling in a single quantum dot-semiconductor microcavity system. Nature **432**, 197–200 (2004)
11. A. Laucht, S. Pütz, T. Günthner, N. Hauke, R. Saive, S. Frédérick, M. Bichler, M.-C. Amann, A. Holleitner, M. Kaniber, J. Finley, A waveguide-coupled on-chip single-photon source. Phys. Rev. X **2**, 1–7 (2012)

12. A. Schwagmann, S. Kalliakos, I. Farrer, J.P. Griffiths, G.A.C. Jones, D.A. Ritchie, A.J. Shields, On-chip single photon emission from an integrated semiconductor quantum dot into a photonic crystal waveguide. Appl. Phys. Lett. **99**(26), 261108 (2011)
13. D. Press, T.D. Ladd, B. Zhang, Y. Yamamoto, Complete quantum control of a single quantum dot spin using ultrafast optical pulses. Nature **456**, 218–221 (2008)
14. T. Godden, J. Quilter, A. Ramsay, Y.Wu, P. Brereton, S. Boyle, I. Luxmoore, J. Puebla-Nunez, A. Fox, M. Skolnick, Coherent optical control of the spin of a single hole in an InAs/GaAs quantum dot. Phys. Rev. Lett. **108**, 1–5 (2012)
15. J.M. Lourtioz, H. Bensity, V. Berger, J.M. Gérard, D. Maystre, A. Tchelnokov, *Photonic crystals: Toward Nanoscale Photonic Devices* (Springer, Berlin, 2003)
16. A. Tandaechanurat, S. Ishida, K. Aoki, D. Guimard, M. Nomura, S. Iwamoto, Y. Arakawa, Demonstration of high-Q (>8600) three-dimensional photonic crystal nanocavity embedding quantum dots. Appl. Phys. Lett. **94**(17), 171115 (2009)
17. J.D. Joannopoulos, S.G. Johnson, J.N. Winn, R.D. Meade, *Photonic Crystals: Molding the Flow of Light* (Princeton University Press, Princeton, 2008)
18. S. Johnson, J. Joannopoulos, Block-iterative frequency-domain methods for Maxwell's equations in a planewave basis. Opt. Express **8**, 173 (Jan. 2001)
19. S. Noda, A. Chutinan, M. Imada, Trapping and emission of photons by a single defect in a photonic bandgap structure. Nature **407**(6804), 608–610 (2000)
20. A.R.A. Chalcraft, S. Lam, D. OBrien, T.F. Krauss, M. Sahin, D. Szymanski, D. Sanvitto, R. Oulton, M.S. Skolnick, A.M. Fox, D.M. Whittaker, H.-Y. Liu, M. Hopkinson, Mode structure of the L3 photonic crystal cavity. Appl. Phys. Lett. **90**(24), 241117 (2007)
21. M. Shirane, S. Kono, J. Ushida, S. Ohkouchi, N. Ikeda, Y. Sugimoto, A. Tomita, Mode identification of high-quality-factor single-defect nanocavities in quantum dot-embedded photonic crystals. J. Appl. Phys. **101**(7), 073107 (2007)
22. T. Yoshie, J. Vučković, A. Scherer, H. Chen, D. Deppe, High quality two-dimensional photonic crystal slab cavities. Appl. Phys. Lett. **79**(26), 4289 (2001)
23. Y. Akahane, T. Asano, B. Song, S. Noda, High-Q photonic nanocavity in a two-dimensional photonic crystal. Nature **425**(October), 944–47 (2003)
24. M. Notomi, K. Yamada, A. Shinya, J. Takahashi, C. Takahashi, I. Yokohama, Extremely large group-velocity dispersion of line-defect waveguides in photonic crystal slabs. Phys. Rev. Lett. **87**, 1–4 (2001)
25. A.Y. Petrov, M. Eich, Zero dispersion at small group velocities in photonic crystal waveguides. Appl. Phys. Lett. **85**(21), 4866 (2004)
26. Y.A. Vlasov, M. O'Boyle, H.F. Hamann, S.J. McNab, Active control of slow light on a chip with photonic crystal waveguides. Nature **438**, 65–69 (2005)
27. T. Baba, Slow light in photonic crystals. Nat. Photon. **2**, 465–473 (2008)
28. M. Bayer, G. Ortner, O. Stern, A. Kuther, A. Gorbunov, A. Forchel, P. Hawrylak, S. Fafard, K. Hinzer, T. Reinecke, S. Walck, J. Reithmaier, F. Klopf, F. Schäfer, Fine structure of neutral and charged excitons in self-assembled In(Ga)As/(Al)GaAs quantum dots. Phys. Rev. B **65**, 1–23 (2002)
29. E. Poem, O. Kenneth, Y. Kodriano, Y. Benny, S. Khatsevich, J. Avron, D. Gershoni, Optically induced rotation of an exciton spin in a semiconductor quantum dot. Phys. Rev. Lett. **107**, 087401 (2011)
30. M. Kroutvar, Y. Ducommun, D. Heiss, M. Bichler, Optically programmable electron spin memory using semiconductor quantum dots. Nature **432**(November), 81–84 (2004)
31. E. Chekhovich, A. Krysa, M. Skolnick, A. Tartakovskii, Direct measurement of the hole-nuclear spin interaction in single InP/GaInP quantum dots using photoluminescence spectroscopy. Phys. Rev. Lett. **106**, 027402 (2011)
32. A. Greilich, S.G. Carter, D. Kim, A.S. Bracker, D. Gammon, Optical control of one and two hole spins in interacting quantum dots. Nat. Photon. **5**, 702–708 (2011)
33. A.M. Fox, *Quantum Optics: An Introduction* (Oxford University Press, Oxford, 2006)

Chapter 2
Experimental Methods

2.1 Introduction

This chapter contains details of the samples, spectroscopic techniques and apparatus used to perform the measurements presented in the subsequent chapters.

It contains information on the growth of quantum dots, and the processes used to fabricate the devices that are studied in this thesis.

General principles of common spectroscopic techniques used to conduct the subsequent measurements are presented alongside details of how they are implemented in the laboratory.

Specific effort is made to detail the experimental set-up used in Chaps. 4 and 5. This is a bespoke system that provides multiple spatially selective excitation and collection paths, integrated with an interferometer and two-colour selection.

2.2 Sample Fabrication

2.2.1 QD Growth

Self-assembled quantum dots are grown by epitaxial growth techniques (A description of molecular beam epitaxy (MBE) and metallic organic vapour phase epitaxy (MOVPE) can be found in [1]). During epitaxial growth a crystal of lower band-gap material is grown over an existing crystal surface of higher band-gap material and is subsequently capped with the higher-band gap material. Current techniques are capable of depositing high quality semiconductor material with monolayer accuracy. This allows abrupt changes in material composition to be achieved. The control of material deposition provided by epitaxial growth is used to produce quantum dots through self assembly. This is the spontaneous formation of 'islands' in the deposited material.

N. A. Wasley, *Nano-photonics in III–V Semiconductors for Integrated Quantum Optical Circuits*, Springer Theses, DOI: 10.1007/978-3-319-01514-9_2,

The formation of the QDs is a result of the strain effects at the interface of the two semiconductor materials, caused by a difference in their lattice constants. The strain can release through two processes depending on the thickness of the material. For a thin mono-layer of material below a critical thickness the deposited material adopts the lattice constant of the host material, for a thicker layer above the critical thickness the deposited layer partially recovers its original lattice constant by releasing the strain through dislocations and defects at the interface.

The self assembled formation of quantum dots occurs as the deposited material crosses the transition between thin monolayers of material and the thick layer of material. The control of the material deposition to monolayer accuracy allows this transition to be used to form miniature islands of material. For a lattice mismatch of 7 % between GaAs and InAs, the growth of InAs cannot be sustained for more than approximately two monolayers (wetting layer) before this transition occurs. The material deposition is stopped before the islands grow into a complete layer. They are then capped with the higher bandgap material, protecting them from the surrounding environment, forming buried QDs.

The QDs studied in this thesis were grown at the EPSRC National Centre for III-V Technologies, University of Sheffield.

2.2.2 Device Fabrication

The devices studied in this thesis are all fabricated using top-down approaches that rely on lithographic, etching and evaporation techniques. To achieve this requires specific designs of the material layers deposited during epitaxial growth that vary depending on the function of the device. For samples that require an under-etch, a sacrificial layer of AlGaAs is included during growth deposition.

Photonic Crystal and Nano-Wire Air-Bridge Structures

A typical wafer design for a GaAs/InGaAs photonic crystal or nano-wire sample as used in Chaps. 3–5 and 7. Section 7.4 is schematically depicted in Fig. 2.1a. (In this

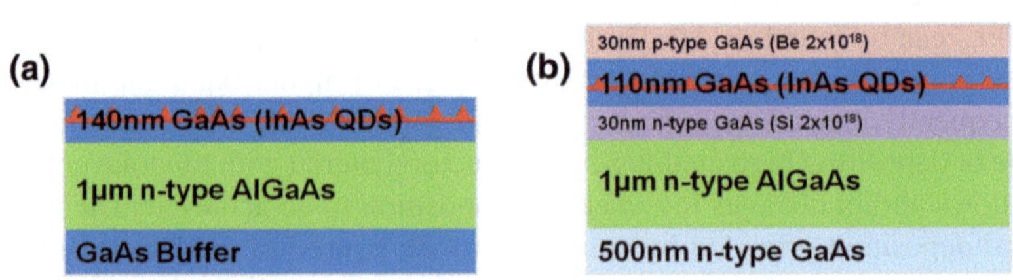

Fig. 2.1 Schematic of the wafer designs used for **a** GaAs/InGaAs photonic crystal or nano-wire samples and **b** GaAs/InGaAs P-I-N diode devices

thesis the term nano-wire refers to a lithographically defined, underetched waveguide of semiconductor material used for in plane light propagation, in contrast to the epitaxially grown pillars of semiconductor material that are orientated perpendicular to a semiconductor substrate as presented in Refs.[2, 3]). It consists of a 1 μm n-type AlGaAs sacrificial layer grown on an undoped GaAs substrate, onto which 140 nm of GaAs is deposited, inside which a layer of InAs QDs are grown. A similar process is used to fabricate all under-etched air-bridge type structures such as photonic crystal cavities, waveguides and nano-wires.

The sample, Fig. 2.2a, is first spun with ZEP 520 A resist (Fig. 2.2b). The pattern (for example a photonic crystal cavity) is then defined in the resist by exposing the sample using electron-beam (E-beam) lithography (Fig. 2.2c). E-beam lithography has a greater resolution than optical lithography which is necessary to pattern the <100 nm features of a photonic crystal.

Once the resist is developed using a Xylene developer (Fig. 2.2d), the pattern is transferred into the semiconductor heterostructure by a reactive ion etch (RIE), Fig. 2.2e. The etch depth is then checked using a dektak to confirm that the etch has penetrated the sacrificial AlGaAs layer, if not, further etching using the RIE may take place until the correct etch depth is achieved. The ZEP 520 A resist can then be removed by heating the sample in n-Methyl-pyrrolidone. Finally, the sacrificial AlGaAs layer is selectively wet etched using hydrofluoric acid (HF), to leave a free standing air-bridge photonic crystal cavity or crossed waveguide structure (Fig. 2.2f).

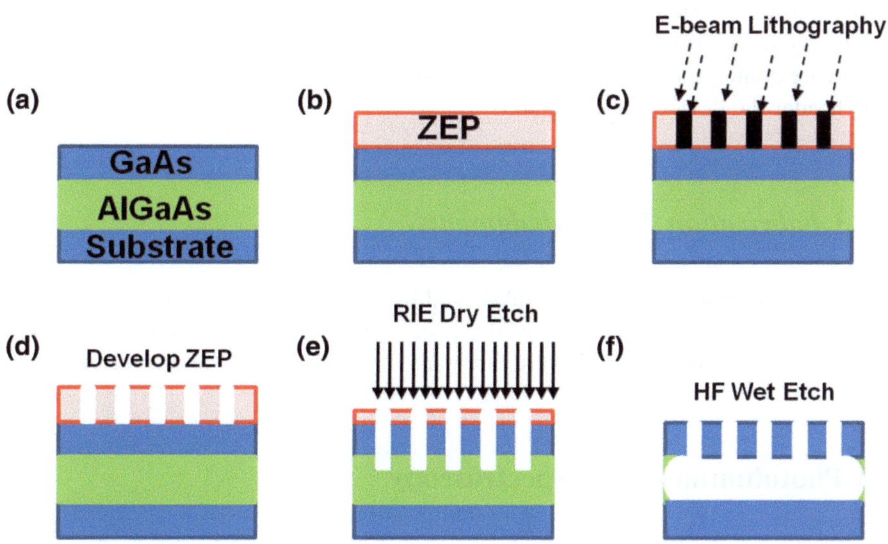

Fig. 2.2 Schematic of the fabrication process used to create air-bridged photonic crystal cavities and similar under-etched structures. **a** A sample containing an AlGaAs sacrificial layer, **b** ZEP 520 A positive resist is spun onto the sample. **c** The pattern is written into the resist by exposure using E-beam lithography. **d** The pattern is developed using a Xylene developer. **e** The pattern is transferred to the sample by etching down to the sacrificial layer using an RIE. **f** Sacrificial AlGaAs is selectively wet etched with HF, leaving the under-etched photonic crystal structure

P-I-N Diodes

For the P-I-N diode devices investigated in Chap. 7, Sect. 7.3, the fabrication process is different from the fabrication process for the simple under-etched photonic crystals which is described above. A typical wafer design for a GaAs/InGaAs P-I-N diode devices, as used in Chap. 7. Section 7.3 is schematically depicted in Fig. 2.1b.

The sample is first spun with BPRS100 resist and a mesa pattern is defined by exposing the resist to UV light through a pre-patterned mask. After the resist is developed the mesa is etched down to the first n-type GaAs layer using a 1:8:80 (Sulphuric acid : Hydrogen peroxide : De-ionised (Di) water) wet etch. Before the resist is removed, the etch depth is checked using the dektak and the wet etch repeated until the correct depth has been achieved.

Once the mesa has been etched, the photonic crystals are then patterned as described above but the final HF under-etching stage is not completed. First, The n-type bottom contact is deposited by spinning the sample with BPRS100 resist and exposing it to UV light through a second pre-patterned 'back contact' mask and developed. Once surface oxides have been removed by immersing the sample in 19:1 (Di water:Ammonia) solution, an InGe:Au (30:200 nm) contact is evaporated onto the sample. The sample is then annealed using a rapid thermal annealer (RTP) at 360 °C for 3 s.

The p-type top contact is deposited by following the same method except the 'back contact' mask is now replaced with the 'top contact' mask and the evaporation uses Ti:Au (20:200 nm). No annealing takes place.

The photonic crystal under-etch can then be completed using the HF wet etch. Once wire contacts are bonded to the top and bottom contacts using gold wire and a TO5 header the device is functional.

2.2.3 *Fabrication Acknowledgments*

The devices used in Chap. 3 were fabricated by I. J. Luxmoore and R. J. Coles. The devices used in Chaps. 4 and 5 were fabricated by I. J. Luxmoore and S. Kasture. The devices used in Chaps. 6 and 7 were fabricated by N. A. Wasley and I. J. Luxmoore.

2.3 Photoluminescence Spectroscopy

Quantum dots can be probed both electrically (electroluminescence) and optically (photoluminescence (PL)). Photoluminescence is the process in which a quantum dot is triggered into emitting light when incident photons promote an electron from the valence band to the conduction band of the QD. This process creates a bound electron-hole pair that forms an exciton. All of the measurements in this thesis have been conducted using optical probing of QDs.

There are two distinct methods of optical excitation for the purpose of optical spectroscopy. Non-resonant excitation uses a laser with an energy that is larger than the energy of the QD transition being observed. In this process the laser creates an excitonic population in higher energy QD states or in the bulk semiconductor material. The higher energy states then non-radiatively relax into the lowest energy QD state, through carrier scattering and phonon interactions. The second approach is to use a laser with energy equal to the excitonic state of interest, so called resonant excitation. In both techniques the electron subsequently decays back into the valence band radiatively recombining with the bound hole to emit a photon.

The optical spectroscopy performed in this thesis was conducted with non-resonant excitation.

2.3.1 Non-resonant Photoluminescence Spectroscopy

Although a quantum dot has discrete electronic energy levels, the surrounding semiconductor material is a continuum of filled valence and available conduction band states. For non resonant excitation an incident laser generates electron - hole pairs in the host semiconductor close to the quantum dot. These pairs may recombine quickly to generate photons at the bandgap energy of the semiconductor, or the electrons and holes may be captured into the discrete exciton levels of the QD.

For this to happen, the excess energy of the excitons must be removed by carrier scattering or phonon interactions. This relaxation process happens relatively fast and removes any coherence with the excitation laser. This process of non-resonant excitation and subsequent radiative recombination from a QD is schematically depicted in Fig. 2.3a. Radiative recombination of an electron-hole pair in the quantum dot consequently reveals information on the quantised energy levels and the optical selection rules.

A typical μPL set-up consists of an incident laser directed by a beam-splitter (BS) to an objective which focuses the light onto the sample. Photoluminescence emitted from the sample is collected by the same objective, passed back through the beam-splitter and directed to a detector where the signal is recorded. A schematic of this excitation geometry is displayed in Fig. 2.3b.

A spectrometer that is able to resolve the spectral content of the emission is often used to disperse the PL prior to detection. Typically our measurements use a 0.55 m single grating spectrometer to disperse the PL which is subsequently detected with a liquid nitrogen cooled charged coupled device (CCD) to record the emission spectrum of the quantum dot.

From the emission spectra recorded on the CCD the neutral exciton, X^0, charged exciton, X^+ and biexciton, XX^0, emission lines, can be identified as well as study of the fine structure of the QD. It is also possible to use this set-up to observe strong coupling in photonic structures and mixing of bright and dark exciton complexes [4].

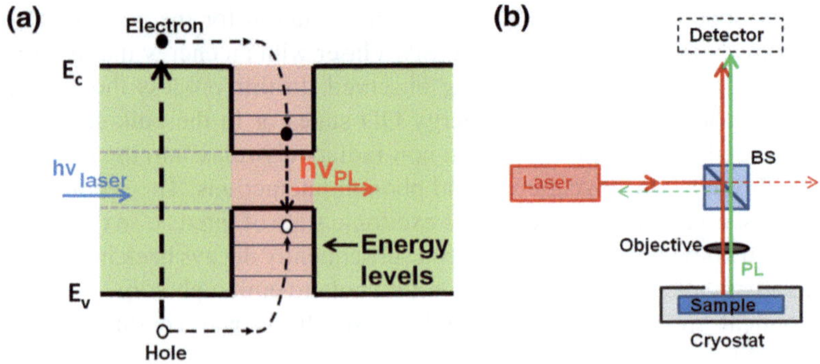

Fig. 2.3 **a** A schematic diagram of the non-resonant excitation process for photoluminescence spectroscopy of QDs. An incident photon of energy $h\nu_{laser}$ greater than the bad gap energy of the host semiconductor E_c–E_v, promotes an electron from the conduction band to the valence band leaving a hole. The hole and electron non-radiatively scatter into the quantised energy levels of the QD where they radiatively recombine emitting a photon of energy $h\nu_{PL}$. **b** Schematic diagram of the experimental excitation geometry for photoluminescence experiments

When optically exciting quantum dots within photonic crystal cavities it is possible to preferentially excite QDs within the photonic crystal cavity by exciting the sample with a laser tuned to a cavity mode [5].

2.4 Cryostats

All of the measurements presented in this thesis were conducted at low base temperatures of $T < 10$ K. To achieve this the samples were housed within cryostat systems, that cool the samples from a source of liquid helium. There are two designs of cryostat used in measurements presented in this thesis. The design and operation of the cryostats are briefly described here.

2.4.1 Cold Finger Continuous Flow Helium Cryostat

The cold finger continuous flow helium cryostat (Oxford Instruments), was used to conduct the majority of experiments in Chaps. 6 and 7. It consist of a small evacuated chamber located on the optical bench in which the sample is housed. The sample is located in a fixed position on top of a copper finger and under a window in the chamber wall which allows optical access to the sample. The copper finger provides a thermal contact between the evacuated chamber and a continuous flow of helium, pumped from a dewar of liquid helium. This system can typically cool a sample to around 10 K. A schematic diagram of a cold finger cryostat is presented in Fig. 2.4.

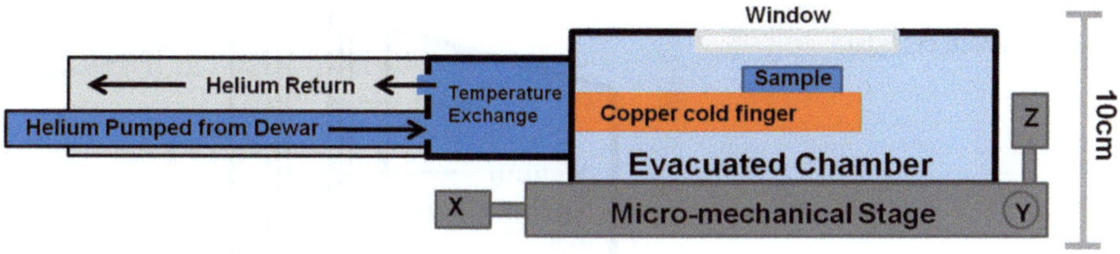

Fig. 2.4 Schematic diagram of a cold finger continuous flow helium cryostat

Due to the vibrations caused by the pumping of liquid helium from the dewar and the constant flow across the cold finger, the stability is somewhat compromised. A laser aligned to the sample will maintain coupling for at most a couple of hours.

In this cryostat, positioning of the laser on the sample is controlled by a a set of micro-mechanical stages which sit beneath the cryostat and on top of the optical bench providing control in x, y and z directions. Fine control is also provided by a piezo-stage mounted objective lens.

2.4.2 Helium Bath Cryostat

A helium bath cryostat is a more stable system than a helium flow cryostat because there is no flow of helium gas. Stable coupling can be maintained over several days. It also reaches temperatures down to 4.2 K below that of the continuous flow cryostat.

The helium bath cryostat, consists of an evacuated jacket containing a bath of liquid helium. The sample and objective lens are mounted inside an evacuated sleeve that sits within the helium bath, and inside the bore of a superconducting magnet. A schematic diagram of a helium bath cryostat is presented in Fig. 2.5.

The superconducting magnet is capable of providing magnetic fields of between $0 < |B| < 5T$. The sleeve contains a small amount of helium exchange gas, to provide good thermal contact with the helium bath. A stack of electrically contacted piezo-electric transducers located within the sleeve positioned the sample when inside the cryostat. The whole system is suspended and insulated to dampen vibrations and improve the stability.

A helium bath cryostat (Attocube Systems) was used to perform the majority of measurements in Chaps. 3–5.

2.5 Time Resolved Measurements

Although photoluminescence spectroscopy as described in Sect. 2.3.1 is able to determine the transition energies between optically active quantum dot states, it cannot associate a timescale with the transitions.

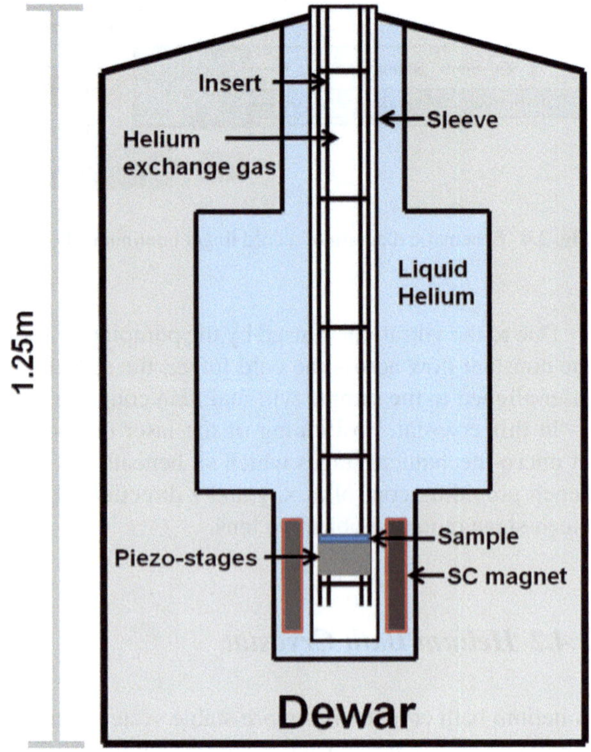

Fig. 2.5 Schematic diagram of a helium bath cryostat

To achieve temporal resolution from the optical spectroscopy of quantum dots electron-hole pairs are generated in the surrounding semiconductor using short non-resonant laser pulses typically a few hundred femtosecond to a few picosecond duration.

Due to QD capture, electron-hole lifetimes in the bulk semiconductor, are on the order of a few picoseconds. Therefore only the excitons captured by the quantum dot remain even after just tens of picoseconds following the initial excitation pulse.

The quantum dot exciton subsequently decays in the absence of any other incident laser excitation, and free from excited carrier induced effects. By spectrally filtering the QD emission from other transitions, a histogram of photon arrival time detected by photon counting photodetectors with respect to each excitation laser pulse can be recorded.

The histogram represents the probability as a function of time that the exciton remains in its excited state after the initial laser pulse. By fitting this trace to an exponential decay a typical lifetime of the QD transition can be extracted. An ideal two-level system will produce a single exponential decay directly related to the excited state lifetime, and thus the natural linewidth of the transition through a Fourier transform.

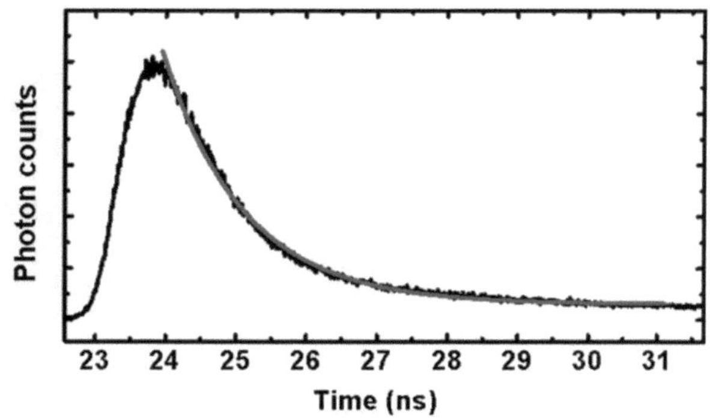

Fig. 2.6 Decay transient of detection events recorded on an APD as a function of time for a single InAs QD excited with a pulsed laser. The *decay curve* is fitted with a single exponential decay that reveals the lifetime of the QD to be 1.2 ns

However, not only the radiative decay, but also any other process that results in exciton population loss affects the decay rate. For example, if there is a non-radiative channel for exciton decay that is faster than the radiative lifetime, the measured decay transient may exhibit a double exponential where the decay of each exponential is dictated by the radiative and non-radiative decay rates.

Figure 2.6 presents data from a single InAs QD measured as described in this section. The decay transient of recorded detection event times is fitted with an exponential decay that reveals the lifetime of the state to be 1.2 ns.

2.6 Second-order Correlation Measurements

For a number of applications in quantum optics a single photon source is a requirement. An ideal single photon source will emit not more than one photon at any given time. This is the case for an ideal two level quantum emitter.

A second-order correlation measurement records the time correlation of an optical signal, revealing the extent to which a QD approximates a two-level emitter.

To perform a second-order correlation measurement on a single QD emission line, the Hanbury-Brown and Twiss (HBT) experimental set-up is used, depicted schematically in Fig. 2.7a.

The collected QD emission is divided by a beam-splitter to be detected by two photon counting detectors (APDs). Photon detection times are then registered at each detector using a single photon counting module (SPCM). A count is recorded conditioned on the previous detection of a photon; this is a coincidence count. A histogram of counts is plotted as a function of the delay time between successive detection events.

The function $g^{(2)}(\tau) = (1 - (1 - M)e^{(-|\tau|/\tau_d)})$ where M is the multi-photon emission probability and τ_d is the lifetime of the QD, can be fitted to the histogram of coincidence, as displayed in Fig. 2.7b.

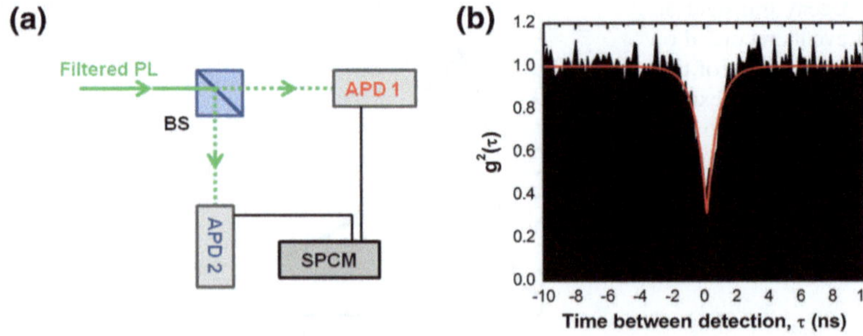

Fig. 2.7 **a** Schematic diagram of a Hanbury-Brown and Twiss experimental set-up. Spectrally filtered PL is incident on a beam-splitter (BS) and directed to two avalanche photo diodes (APD) that record incident photons and pass a signal to the single photon counting module (SPCM). The SPCM is responsible for recording a count conditioned on a detection of a previous photon detection signal. **b** An example of the coincident count histogram of detection events recorded by the SPCM for a single InAs QD transition. The function $g^{(2)}(\tau) = (1-(1-M)e^{(-|\tau|/\tau_d)})$ is fitted (*red line*) revealing a multi photon probability, $M = 0.28$ and and lifetime of the QD $\tau_d = 0.76$ ns

For a coherent light source (such as laser emission) the histogram exhibits no correlation and the multi photon probability, $M = 1$, such that $g^{(2)}(\tau) = 1$ for all delay times. The single photon emission from a two level system results in anti-correlation $M < 1$. An ideal system will never record two photon counts at once and $M = 0$, and the central 'dip' will go to zero.

The anti-correlation of photon emission is also known as photon anti-bunching.

Photon correlation measurements can also reveal more about the internal dynamics of multi-exciton recombination by recording the intensity correlations between any two QD emission lines. These measurements exploit the well distinguished energy shifts of individual recombination between neutral and charged excitons as well as biexcitons to channel signal to the individual APDs by spectral filtering, rather than using the beam-splitter described above. Similarly, this technique can be used to determine whether two transitions observed from different locations originate from the same single source, as implemented in Chap. 5.

2.7 Two Colour, Spatially Selective Collection and Interference Measurements

The non-resonant spectroscopy, time-resolved and second order correlation measurements described above are standard tools in the optical spectroscopy of quantum dots and are performed using standard experimental set-ups widely used in many optical spectroscopy labs. As such, the details of the experimental set-ups are only briefly discussed in the relevant chapters where any functionally significant variations are detailed.

In contrast, the experimental set-up used in Chap. 4 to demonstrate in-plane polarisation transfer required a design and development process that integrated and modified optical techniques such as interferometry with bespoke solutions for multiple spatially selective and independently controlled collection paths. Thus the details of the experimental set-up are detailed here.

2.7.1 Spatial Selection

Key to the demonstration of in-plane polarisation transfer presented in Chap. 4 is the ability to excite and collect from three spatially separate locations.

The excitation path for the laser is the same as for a standard optical spectroscopy set-up, schematically depicted in Fig. 2.3. The laser path is controlled with mirrors (CM) and reflected onto the sample by a beam-splitter (BS).

Typically the collection path in a optical spectroscopy experiment is defined by the position of the excitation laser, whereby the path of the reflected laser from the sample approximately defines the collection path of PL. Thus PL is collected from the position of excitation.

To provide spatially selective collection with positioning that can be controlled independently, single mode fibers (SMF) are used in a manner similar to a pin-hole in the collection path. Only light from a spatially defined part of the field of view couples into the single mode fiber. By employing a second beam splitter in the collection path, a second arm of PL collection is formed which includes a second SMF coupler providing a second spatially selective collection spot. The single mode fibers used in the experiment were designed for a wavelength operation of (830 ± 150) nm, with a 5.6 μm mode field diameter (MFD) and a 125 μm cladding.

In the experiments conducted in Chap. 4, the emission from orthogonal outcouplers is orthogonally polarised. This allows a polarising beam-splitter (PBS) to be used as the second beam-splitter reducing the signal loss. The PBS could itself be replaced with a mirror positioned to reflect a single path if required, as the function of the PBS is to provide two spatially independent collection paths and not to divide a single path. When a PBS is in-place a 1/2 waveplate (1/2 WP) located prior to the PBS allows the orthogonality of the PL emission to be aligned with the polarisation axis of the PBS to minimise signal loss.

The size of the collection spot provided by the single mode fiber can be controlled through a combination of focal lengths of the sample objective and the the coupling lens into the fiber, as well as the radius of the optical fiber. In order to achieve good coupling efficiency, the PL beam is matched to the field diameter of the fiber mode. This is done by choosing a coupling lens with the appropriate numerical aperture (NA). This provided us with a collection spot of approximately 2 μm in diameter and maximum coupling efficiency into the single mode fiber of $\sim$50 % as determined by tests that coupled laser light directly into the SMFs.

For the spatially selective excitation and collection used in Chap. 3 to investigate the PCWs, a similar system was used. However, to improve the collection efficiency

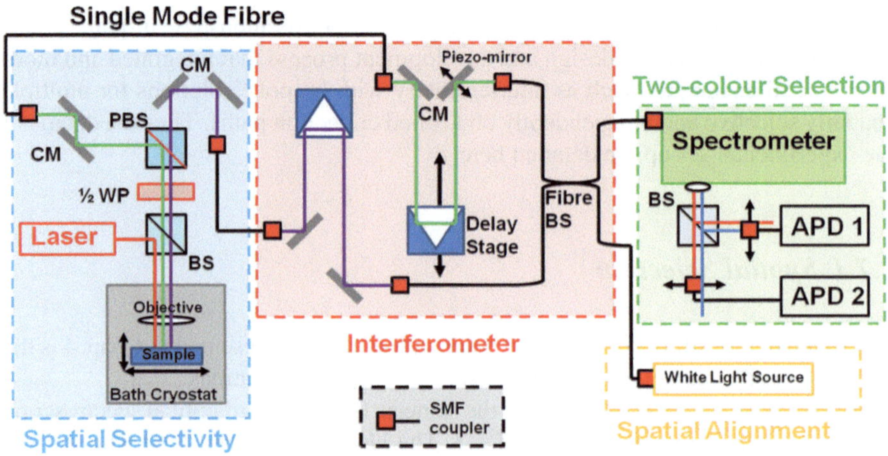

Fig. 2.8 Schematic diagram of the experimental set-up used to provide two-colour, spatially selective collection and interference measurements. The four subsections of the functionality required in the measurement are highlighted with coloured boxes for visual reference and to emphasise that the subsections can be individually incorporated into other experimental set-ups as required

only one collection path was used and the SMF was replaced with a multimode fiber (MMF).

The experimental apparatus involved in the spatial selection process highlighted in red in Fig. 2.8, was physically located on a 'breadboard' system positioned on top of a helium bath cryostat. The optical connection between a 'breadboard' and the apparatus located on the optical bench was provided by the SMFs.

2.7.2 *Spatial Alignment*

The positioning and alignment of the spatially selective collection paths provided by the SMFs is conducted using a combination of control mirrors (CM), high precision manual stages and a white light source.

By coupling a white light source into the single mode fiber in the opposite direction to the collection of PL emission, as shown in the yellow box of Fig. 2.8, a highlighted image is formed on the sample corresponding to the spatial selection provided by the SMFs. By imaging the sample with a camera these areas are visible on a TV screen.

The position from which PL is collected can then be manipulated with the control mirrors (CM) in the collection path and high precision manual stages on which the SMF couplers are mounted. This technique provides a visible guide to the position and size of the collection spot. The size of the collection spot can be tuned using the focus of the SMF coupler lens.

2.7.3 Interferometer

Once the spatially selective collection into the SMFs has been optimised, the output of the SMF enters the interferometer section of the set-up shown schematically by the red box in Fig. 2.8. It is important to note that the use of SMFs is critical if interferometry of the two collection paths is to be conducted. The multiple optical paths available through multi mode fibers (MMF) scrambles the signal coherence.

To form an interferometer one collection path includes a controllable free-space delay path. This is provided by a combination of a mirror mounted motorised stage and a piezo-actuated mirror. The motorised delay stage provides coarse delays of approximately picosecond resolution, whilst the piezo-actuated mirror allows the path delay to be controlled over a few wavelengths.

So that the optical path length within the fibers remain equal, the second path has a fixed free-space path delay approximately equal in length to the first.

After the free-space sections, the optical path is coupled into a SMF 50:50 beam-splitter. This beam-splitter interferes the signal from the two collection paths for the first time.

The combined signal then passes from one out-port of the beam-splitter into a 0.55 m single grating spectrometer. The spectrometer is used to disperse the PL and filter regions of spectral interest from scattered laser signal and unwanted PL.

2.7.4 Two-colour Selection

An important feature of the experiments conducted in Chap. 4 is the ability to perform interference experiments simultaneously for two wavelengths separated in wavelength by around ~0.4 nm.

The post-spectrometer two-colour selection set-up is displayed in the green box of Fig. 2.8. The spectrometer and the side exit port are used to spectrally filter and disperse the spectral region of interest. The dispersed spectral region of PL exits the spectrometer and is incident on a beam splitter creating two optical paths.

Each optical path is coupled into a single mode fiber. As with the collection paths, the coupling into the SMFs is used to spatially filter a portion of the dispersed PL spectrum. In effect, this selects a subsection of the spectral region that has exited the spectrometer.

The SMF coupler is mounted on a motorised stage with the axis of motion orthogonal to the optical path. This allows the SMF to be scanned across the spectral range providing control over the subsection of PL spectrum being coupled into the SMF. This is repeated in the second optical path, allowing independent selection of two spectral ranges and hence providing the two-colour selection.

Once coupled into the SMFs the signal for each spectral window is recorded on separate APDs.

With the addition of a SPCM module this set-up can be used to perform second-order correlation measurements, as described in Sect. 2.6. By removing the interferometer from the optical path, second-order correlation measurements can be performed on signals collected from two remote positions on the sample, for example, signal emitted from orthogonal out-couplers.

References

1. P.Y. Yu, M. Cardona, *Fundamentals of Semiconductors: Physics and Materials Properties* (Springer, Berlin, 1996)
2. T.M. Babinec, B.J.M. Hausmann, M. Khan, Y. Zhang, J.R. Maze, P.R. Hemmer, M. Loncar, A diamond nanowire single-photon source. Nat. Nanotechnol. **5**, 195–199 (2010)
3. M.N. Makhonin, A.P. Foster, A.B. Krysa, P.W. Fry, D.G. Davies, T. Grange, T. Walther, M.S. Skolnick, L.R. Wilson, Homogeneous array of nanowire-embedded quantum light emitters. Nano Lett. **13**, 861–865 (2013)
4. M. Winger, A. Badolato, K. Hennessy, E. Hu, A. Imamoğlu, Quantum dot spectroscopy using cavity quantum electrodynamics. Phys. Rev. Lett. **101**, 226808 (2008)
5. M. Kaniber, A. Neumann, A. Laucht, M.F. Huck, M. Bichler, M.-C. Amann, J.J. Finley, Efficient and selective cavity-resonant excitation for single photon generation. New J. Phys. **11**(1), 013031 (2009)

Chapter 3
Disorder Limited Photon Propagation and Anderson Localisation in Photonic Crystal Waveguides

3.1 Introduction

Photons are robust carriers of quantum information [1] and there is great potential for their use in a measurement based linear optics approach to quantum information processing [2]. This approach has recently been used to manipulate multiple qubits in a silica-on-silicon waveguide circuit [3]. By using III–V based materials instead, significant benefits in miniaturisation and increased functionality can be realised. In particular, quantum dots (QDs) can be used for integrated single photon sources and non-linear devices [4] that can be incorporated into waveguide designs to provide a scalable route to QIP.

Photonic crystal waveguides (PCWs) are a leading technology for photon transfer in integrated QD circuits [5]. Advantages of PCWs over competing technologies such as photonic nanowires are the efficient coupling of QD emission [6–8] and the capability to engineer the local density of optical states (LDOS) to produce broadband slow light [9], providing the Purcell enhancement of QD spontaneous emission required to achieve indistinguishable photons from InAs QDs [10, 11]. Alternative cavity-based means to obtain large Purcell enhancements exist, but at the price of significantly reduced bandwidth and mode overlap compared to slow-light approaches [9].

A number of successful demonstrations using this PCW technology, including single photon transfer [12] and Purcell enhanced emission [13] have already taken place. However, photon propagation lengths inevitably suffer when operating in the slow-light regime due to enhanced scattering at low group velocities, thus limiting the scalability of slow light based optical circuits [14, 15].

This chapter studies W1 PCWs (single line of missing air holes) of three lengths, terminated at both ends with out-couplers [16] and fabricated from GaAs that incorporates a single layer of InAs QDs. The presence of Fabry–Perot (FP) resonances in the waveguides provides a mechanism to investigate waveguide loss and photon propagation lengths, when operating at low group velocities close to the waveguide band edge [17]. The localisation lengths deduced from FP analysis are found to

N. A. Wasley, *Nano-photonics in III–V Semiconductors for Integrated Quantum Optical Circuits*, Springer Theses, DOI: 10.1007/978-3-319-01514-9_3,

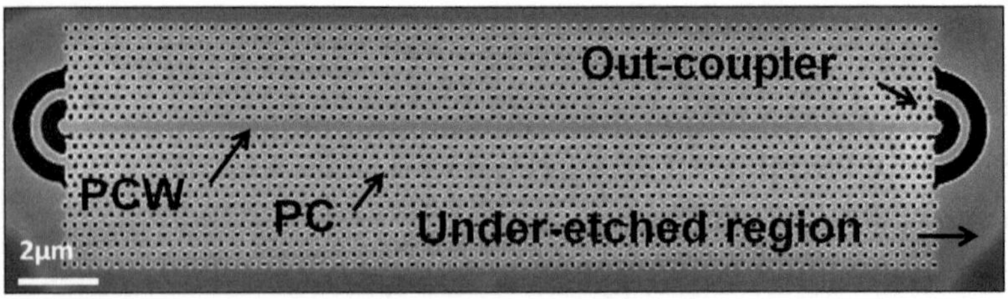

Fig. 3.1 SEM image of a 20 μm W1 PCW end-terminated with out-couplers fabricated from a 130 nm GaAs slab. The lattice constant a = 254 nm and hole radius r = 71 nm

be consistent with direct evidence for localisation observed from the PCW surface, where increased scattering from disorder in the fabricated structures leads to the observation of Anderson-localised modes in the slow light regime [18, 19].

The chapter goes on to discuss the consequences of this work and possible applications for PCWs. In particular the use of PCWs as on-chip phase shifters as well as engineering the PCWs to provide flat-band slow-light, for which preliminary results are presented.

3.2 Sample

The samples used in this study are grown by molecular beam epitaxy (MBE) on undoped GaAs (100) wafers. The wafer consists of a 130 nm GaAs waveguide slab with a single layer of nominally InAs QDs at its centre. This is separated from the GaAs buffer layer by a 1 μm thick $Al_{0.6}Ga_{0.4}As$ sacrificial layer. Electron beam lithography is used to define the PCWs and the GaAs slab layer is etched using an inductively coupled reactive ion etch (ICP-RIE). Finally, hydrofluoric acid is used to selectively remove the $Al_{0.6}Ga_{0.4}As$ layer from beneath the PCWs leaving a free-standing structure. Figure 3.1 shows a scanning electron microscope (SEM) image of a completed PCW. The PCWs have a lattice period, a = 254 nm and hole radius, r = 71 nm. The out-couplers [16] at the ends of the PCW scatter light vertically and allow us to use spatial selective photoluminescence with a single microscope objective to measure the light propagating along a bound PCW mode.

3.2.1 Slow-light Mode

The MIT photonic bands (MPB) [20] software package was used to calculate the band structure for the PCWs and the results are displayed in Fig. 3.2. The guided mode highlighted in red (Mode 1) is the subject of our investigation.

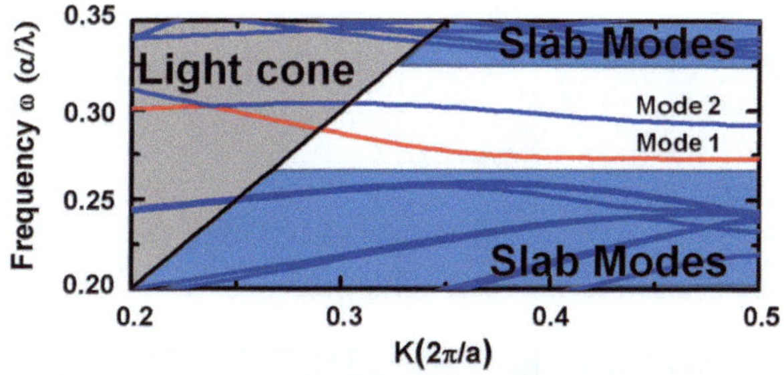

Fig. 3.2 A band diagram calculated using MPB for a W1 GaAs PCW with r/a = 0.28. The mode 1 (*red*) indicates the first guided mode studied in this chapter. The *light line*, $w = ck$ and *light cone* are shaded *grey* whilst other modes not localised to the PCW or outside the photonic bandgap are labelled *slab modes* and shaded *blue*

The slow-light region of the guided mode highlighted in Fig. 3.2 that forms the basis for the investigation presented in this chapter, occurs toward the edge of the Brillouin zone where $d\omega/dk$ falls to zero and the mode looks 'flat'.

3.3 Experimental Geometry

The experimental measurements were performed with the sample held at ~4.2 K in a liquid helium bath cryostat and excited with an 850 nm diode laser focused to a spot of 1 μm by a 0.62 numerical aperture lens. The QD emission was filtered with a 900 nm long pass filter to exclude scattered laser light and then dispersed by a 0.55 m single spectrometer before being recorded with a liquid nitrogen cooled charged coupled device camera (CCD). Spatially selective micro-photoluminescence (μPL) through a single microscope objective provided independent control over the area on the sample from which QDs were excited and PL collected. The spatial selection was achieved by filtering the PL emission using an optical fiber positioned in the image-plane. The PCWs were terminated with out-couplers that scattered propagating light vertically. In this way, PL could be collected from either the point of excitation or the out-coupler whilst exciting along the waveguide (see Fig. 3.3), allowing for the simultaneous study of surface losses and in-plane propagation.

3.4 PCW Characterisation

The propagating mode of the PCWs, was characterised by collecting PL from one of the out-couplers whilst exciting the QD emission at the centre of the structure. A typical PL spectrum is presented in Fig. 3.4 and displays the distinctive Fabry-Perot (FP) type resonances, similar to those seen in other PCW systems [13]. The FP modes are not visible when observing perpendicular to the centre of the waveguide.

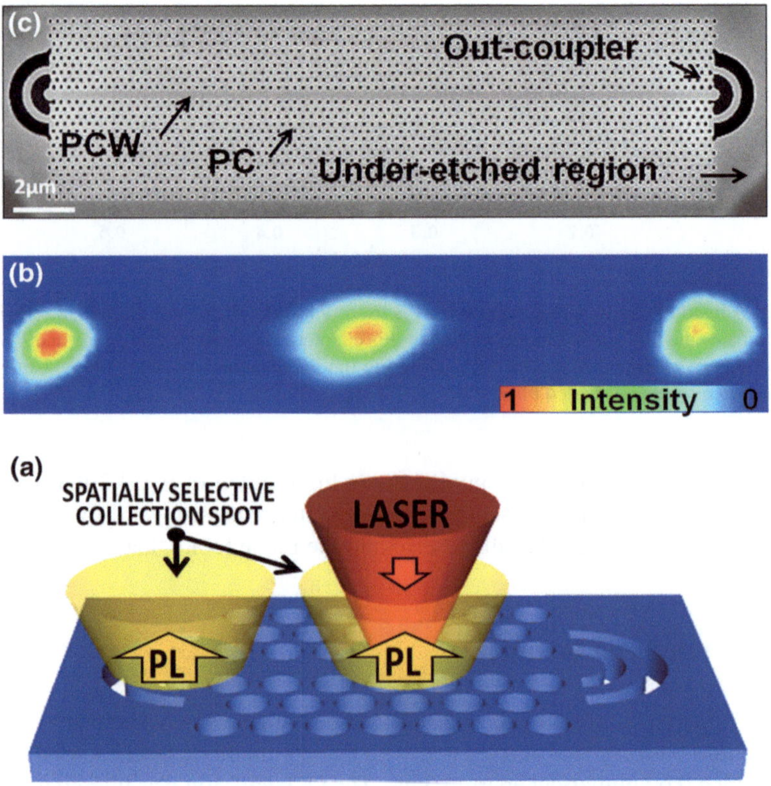

Fig. 3.3 **a** 3D diagram of the experimental set-up with laser excitation spot (*red cone*) and spatially selective PL collection (*yellow cones*) both of which can be independently positioned on the sample (*blue*). **b** PL intensity map of the PCW shown in the SEM diagram, **c** produced by scanning the spatially selective collection spot whilst the centre of the PCW was excited by a 850 nm laser. Three areas of bright PL are observed; at the location of excitation and from the the two outcouplers that terminate the PCW. This demonstrates that PL excited at the centre propagates along the PCW and is scattered vertically by the outcouplers

Typically in such spectra the only observed feature is the broad QD distribution, occasionally additional features are observed and these are discussed in detail in Sect. 3.6.

3.5 Analysis of Fabry–Perot Resonances in PCWs

3.5.1 Group Index

Analysis of the FP modes allows us to extract information about the group index of the guided mode. The group index, n_g, is extracted from Fig. 3.4 using the equation

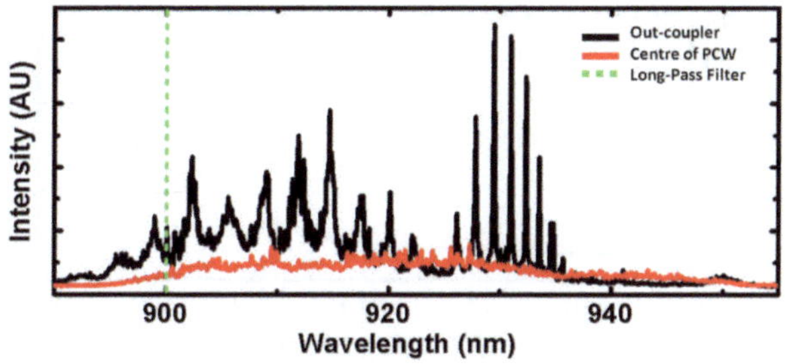

Fig. 3.4 μPL spectra recorded from a 20 μm PCW when laser excitation spot is positioned at the centre of the PCW and the light is collected from an out-coupler (*solid black trace*) and the point of excitation (*red*). The 900 nm long-pass filter edge is superimposed for reference (*green dash*). FP type resonances are observed from the outcoupler but not from the centre of the waveguide, where only emission from the QD ensemble is collected

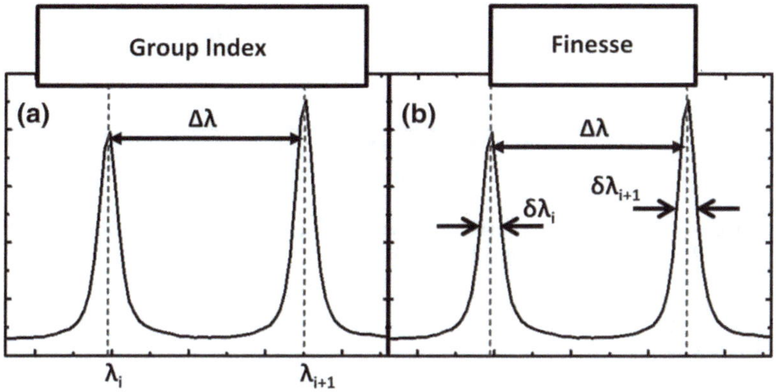

Fig. 3.5 Example of how the properties of the FP fringes are used to calculate **a** the group index of the guided mode and **b** finesse of the FP cavity

$$n_g = \frac{\lambda_i^2}{\Delta\lambda_i 2l} \tag{3.1}$$

where λ is the central wavelength of the ith mode, $\Delta\lambda$ is the free spectral range (the separation between consecutive modes where $\Delta\lambda_i = (\lambda_{i+1} - \lambda_{i-1})/2$ and l is the length of the waveguide, as shown in Fig. 3.5a. The group index is related to the group velocity, v_g, via the equation $v_g = (dw/dk) = c/n_g$. The free spectral range varies with wavelength due to the change in dispersion within the waveguide mode causing a change in group velocity.

The group index for a 20 μm PCW with a $r/a = 0.28$ is plotted as a function of wavelength in Fig. 3.6 and shows a monotonic increase of n_g up to a maximum measured value of 36. The group index for the simulated waveguide mode, calculated from the values of w and k provided by the simulation of the waveguide as displayed in Fig. 3.2, is also plotted in Fig. 3.6 and compares well with the observed data.

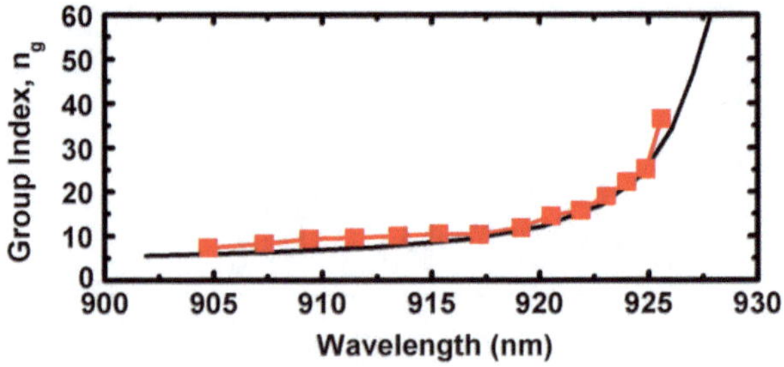

Fig. 3.6 Comparison of group index extracted from analysis of FP resonances shown in Sect. 3.8 (*red*) and extracted from MPB simulations (*black*)

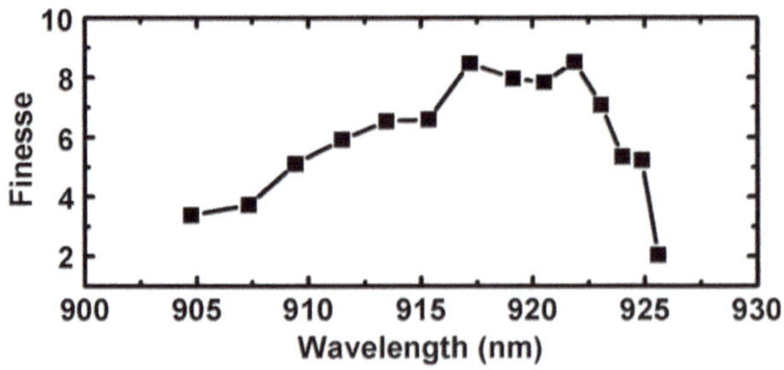

Fig. 3.7 Finesse of the Fabry–Perot (FP) resonances observed from an out-coupler as shown in Fig. 3.8, where $F = \Delta\lambda_i/\delta\lambda_i$ as defined in Fig. 3.5

3.5.2 Finesse

The finesse of the waveguide cavity, F, can be deduced from the width of the FP modes, $\delta\lambda$, according to $F = \Delta\lambda_i/\delta\lambda_i$ and is plotted in Fig. 3.7. It is apparent that the finesse increases with wavelength between 900 and 920 nm before reducing over the final few visible resonances. The initial increase in finesse with group index arises from the increase in the reflectivity of the waveguide facets as the index contrast $(n_g - n_{air})$ increases. The subsequent reduction is due to enhanced loss as the group velocity is reduced, making the effects of disorder more significant.

3.6 Anderson-localisation

It has been explained in Sect. 1 how breaking the periodic dielectric modulation within a photonic crystal can lead to localised states appearing within the photonic bandgap. In practice this can be achieved in a number of ways. Typically a point defect is engineered into a photonic crystal that will have a pre-determined effect on the bandstructure, for example photonic crystal cavities (L3, H1, etc). However

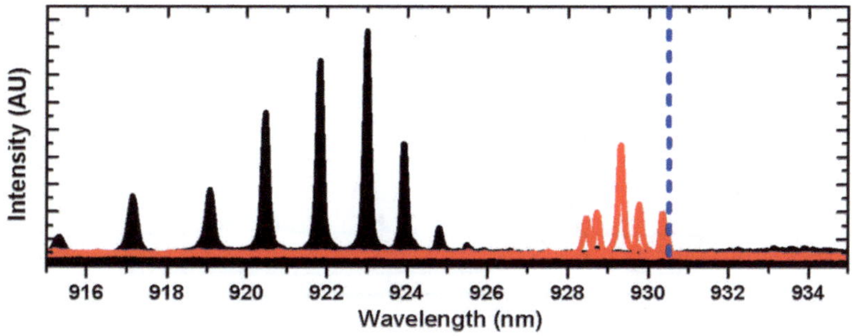

Fig. 3.8 μPL spectra recorded from a 20 μm PCW when the laser excitation spot is positioned at the centre of the PCW and the PL is collected from an out-coupler (*solid black trace*) and when the μPL spectra is recorded for a number of positions above the PCW as the excitation and collection spots are scanned simultaneously along the PCW (*red*). A number of distinctive features appear at localised positions. The waveguide band edge extracted from the simulations is superimposed for reference (*blue dash*)

there is also a probabilistic approach to localisation that relies on random fluctuations within the photonic crystal to confine light. First developed to describe electrons in atomic crystals, the theory behind Anderson-localisation [21] is directly extendable to light propagation in photonic media where there have been many observations, however it was first observed in PCWs by Topolancik et al. in 2007 [18].

The random fluctuations, or 'disorder', in the periodicity of the dielectric is responsible for the multiple scattering of photons leading to wave interference that can result in strong localisation. In the case of PCWs, the disorder can arise through the fluctuations in the position of air-holes relative to the periodic lattice sites or through the shape of the holes. Although this can be achieved by intentionally fabricating photonic crystals with some pre-determined randomisation of hole location or shape, in reality there is always some level of unintentional disorder present in the photonic crystal due to the limit in fabrication accuracy, this can result in the unintentional localisation of propagating light.

Although Anderson-localisation represents an interesting route toward cavity QED [19, 22], the random localisation of propagating photons is an unwanted loss process in a quantum optical circuits that may limit the scalability of these devices.

To gain further insight into the loss occurring in the slow-light regime of our PCWs, PL emitted vertically from the point of excitation is collected. Under these conditions modes are sometimes observed that differ from the FP resonances visible from the out-couplers and in addition to the QD ensemble displayed in Fig. 3.4 (which is an example of a typical PL spectra when collecting in this geometry). These modes occur in a narrow spectral window, for the PCW studied in Sect. 3.5 this is between 928 and 931nm as shown in Fig. 3.8.

By scanning the excitation laser and collection spot along the length of the waveguide we find that the modes occur at irregular locations along the waveguide (Fig. 3.9).

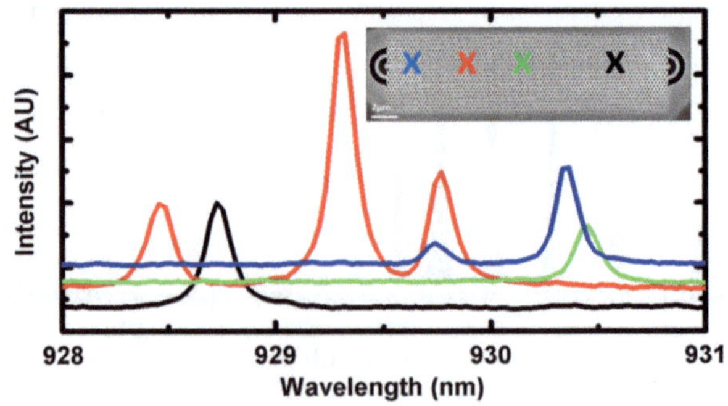

Fig. 3.9 PL spectra of Anderson modes observed at four *colour coded positions* marked on the SEM image of the PCW inset

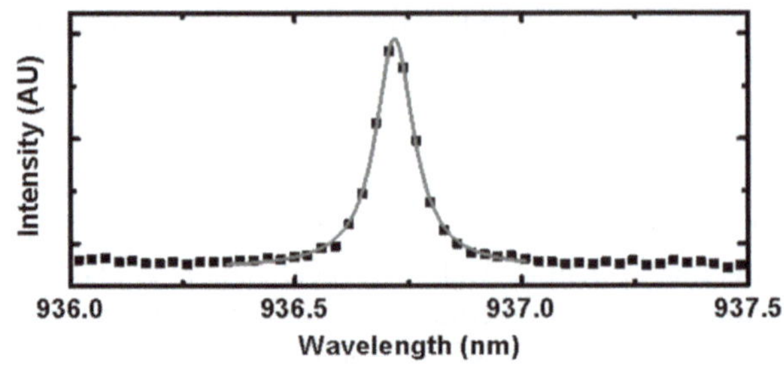

Fig. 3.10 PL spectra of an Anderson mode with a spectrometer resolution limited Q-factor of 9,300

The modes vary both in intensity and linewidth and have Q-factors up to 9,300 limited by spectrometer resolution, as shown in Fig. 3.10.

From the simulations of the band structure the band edge of the mode is calculated to occur at a wavelength of 930.5 nm and identified with the blue dashed line in Fig. 3.8. The modes are spectrally located between the last FP resonance and the band edge, where the simulations indicate $n_g > 50$. We identify these as Anderson-localised modes caused by backscattering due to unintentional disorder present in the PCW [19].

3.7 Scatter at Low Group Velocity

The drop in finesse and presence of Anderson-localised modes indicate substantial waveguide losses occur in the PCWs due to disorder in the photonic crystals. Waveguide loss processes limit the propagation lengths of photons within PCW modes and are therefore of fundamental importance for the design of integrated optical circuits.

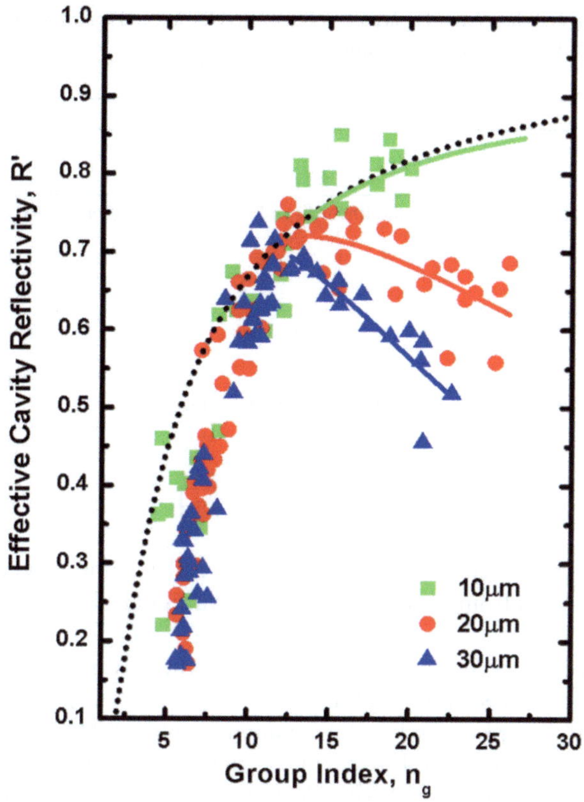

Fig. 3.11 Effective reflectivity, R', for 10 μm (*green squares*), 20 μm (*blue circles*) and 30 μm (*red triangles*) PCWs. Fresnel model of facet reflectivity where $R = ((n_g - n_{air})/(n_g + n_{air}))^2$ (*dashed black line*). $R' = Re^{-\alpha l}$ fitted to the data above $n_g = 12$ for the 20 μm waveguide (*blue line*), the 30 μm waveguide (*red line*) and the 10 μm waveguide (*green line*) where $\alpha = An_g + B$: A = 12(±1) × 10^{-4} nm^{-1}, B = −15(±3) × 10^{-3} nm^{-1}

There are a number of mechanisms responsible for the propagation losses observed in the PCWs. These include out of plane scattering into free space, in plane scattering into the surrounding photonic crystal slab, absorption, backward scattering into the backward propagating mode and inter-mode scattering into other waveguide modes.

The strength of each of these loss processes is dependent on the group velocity of the propagating photons. Out of plane losses scale with a $\frac{1}{v_g}$ relationship while backscattering in the propagating mode has been shown to scale as $\frac{1}{v_g^2}$ at high n_g and is the dominant process in this region [17].

In order to quantify the propagation losses in our PCWs at low group velocities we compared the PL spectra from PCWs of three different lengths; 10 μm, 20 μm and 30 μm.

Table 3.1 Summary of the A and B parameters extracted by fitting Eq. 3.2 with $\alpha = An_g + B$ to the recorded data of R′ for each PCW length.

Parameter	PCW length			Linear fit
	10 μm	20 μm	30 μm	(Fig. 3.12)
A ($\times 10^{-4}$) nm^{-1}	–	11(±1)	16(±2)	12(±2)
B ($\times 10^{-3}$) nm^{-1}	–	−13(±3)	−14(±1)	−15(±3)

The A and B parameters provided by the linear fit in Fig. 3.11 are also presented for comparison

An effective reflectivity

$$R' = Re^{-\alpha l} \tag{3.2}$$

of the lossy-FP cavity can be evaluated from the finesse

$$F = \frac{\pi}{2arcsin\left(\frac{2\sqrt{R'}}{(1-R')}\right)} \tag{3.3}$$

where R is the reflectivity of the waveguide facet and α is the loss coefficient. The values of R′ extracted from the FP resonances in waveguides of differing lengths are plotted in Fig. 3.11.

We first consider the behaviour for $n_g \leq 12$, where R′ is similar for all PCW lengths and R′ increases from ~0.2 at $n_g = 6$ to ~0.75 at $n_g = 12$. The dashed black line in Fig. 3.11 plots the Fresnel equation for reflectivity, R, at a planar facet between two materials with a refractive index contrast, $R = ((n_1 - n_2)/(n_1 + n_2))^2$. For $n_g < 12$, the measured effective reflectivity, R', is less than this simple model predicts. Since R' is similar for the PCWs of differing length, we infer that the waveguide propagation losses are negligible when $n_g < 12$ and the discrepancy between the ideal facet reflectivity and the experimental data is due to the high transmission properties of the out-coupler designed for $n_g = 3.5$ [16].

For $n_g > 12$, R' is dependent on both n_g and the length of the PCW. There is a deviation from R that increases with length, which indicates that waveguide losses begin to degrade the effective cavity reflectivity. For the shortest waveguide, R' lies close to the theoretical value for the planar facet, R. As any waveguide losses degrade the observed value of R' we can assume that the idealised value for the reflectivity is a good approximation to the out-coupler facet reflectivity in this region.

To determine how α scales with n_g in the device and extract information about the propagation losses in the waveguide, $R' = Re^{-\alpha l}$ is fitted to the data for values of $n_g > 12$ with the data rounded to the nearest whole value of refractive index. In this way a near linear variation of α against n_g is deduced for the PCWs, shown in Fig. 3.12. (The specific details of the fit are: $\alpha = An_g + B$ with A $= 12(\pm 2) \times 10^{-4}$ nm^{-1} and B $= -15(\pm 3) \times 10^{-3}$ nm^{-1}).

The same values of $\alpha(n_g)$ may also be used to calculate the overall variation of R' against n_g and are found to be in good agreement with the experimental data as shown in Fig. 3.11.

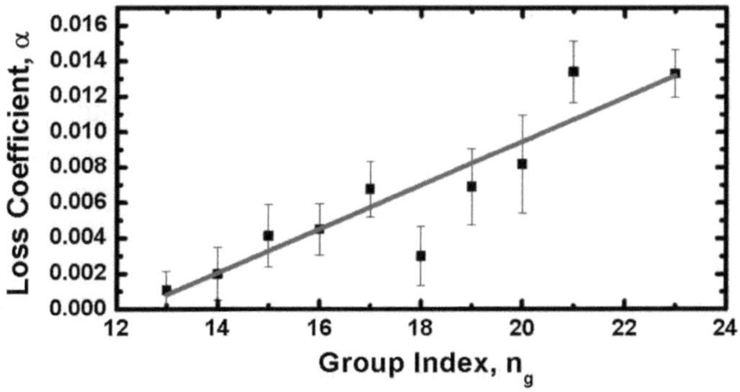

Fig. 3.12 Loss coefficient, α, taken from the data of reflectivity rounded to nearest group index fitted with $R = R'e^{-\alpha l}$ for each value of group index. A linear function, $An_g + B$, is fitted and parameters A = $12(\pm2) \times 10^{-4}$ nm^{-1} and B = $-15(\pm3) \times 10^{-3}$ nm^{-1} extracted

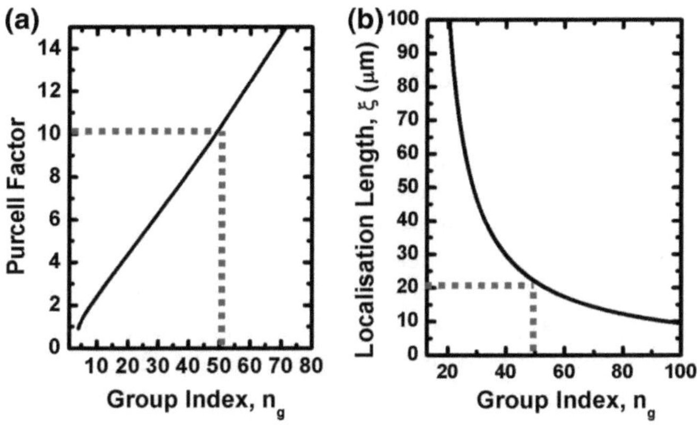

Fig. 3.13 **a** Calculated Purcell factors as a function of group index for the W1 PCW. For a Purcell factor of 10 it is shown that a group index of 50 is required, which is shown to restrict the localisation length to 20 μm, by the *dashed grey line*. **b** Localisation length, ξ extrapolated from the value of α extracted on Fig. 3.12 where $\xi = \frac{ln(1/e)}{\alpha}$

Equation 3.2 with $\alpha = An_g + B$ was also fitted to the data of R' against n_g for the individual waveguide lengths. The values for the A and B parameters extracted in this way for the different length PCWs are presented in Table 3.1 along with the values extracted from Fig. 3.11 for comparison.

It is difficult to extract accurate A and B parameters for the 10 μm sample due to the spread in data points that lie close to the theoretical value of R, therefore there is no data for the 10 μm PCW in Table 3.1. However, the parameters for the 20 and 30 μm waveguides displayed in Table 3.1 are in close agreement with the values taken from the linear fit.

3.8 Localisation Length in PCWs

The propagation length, L, when intensity, I, falls to $\frac{1}{e}$ of the initial value, I_0, given by the relation $I = I_o e^{-\alpha l}$ is $L = \frac{ln(1/e)}{\alpha}$. By assuming the out of plane scattering loss and inelastic material absorption are negligible compared to the disorder induced backscattering loss in a 1D single mode PCW, the propagation length, L, is equal to the localisation length, ξ so that $\xi = \frac{ln(1/e)}{\alpha}$ [23–25].

By extrapolating the value of α given by the fit in Fig. 3.12 an estimate of ξ for n_g up to 100 can be made, as plotted in Fig. 3.13b. The group index corresponding to a localisation length of 20 μm is found to be $n_g = 50$. This is consistent with our experimental observations, where Anderson-localised modes are only observed for $n_g > 50$ in 20 μm PCWs.

3.9 Photon Propagation Limit for Viable Integrated Quantum Optical Circuits

The propagation losses that occur in the PCWs must be taken into consideration when designing integrated optical circuits to operate in the slow-light regime. For the system to be useable as an indistinguishable photon source the QD coherence time, T_2, needs to approach the lifetime limited state where $T_2 \sim 2T_1$ and T_1 is the radiative lifetime of the QD. When we consider for a typical GaAs QD with $T_1 \sim 1ns$ and a $T_2 \sim 400\ ps$ (shorter when located in photonic structures [10, 11]), a Purcell enhancement > 5 is required to achieve this [26, 27].

The Purcell enhancement provided by the PCWs as a function of group index, is modelled as described in Ref. [28] and plotted in Fig. 3.13a. We calculate a Purcell enhancement of 10 can be achieved for $n_g > 50$. However, when working at this group index, the photon propagation length is less than 20 μm, which sets an upper limit to the size of quantum optical circuits constructed from PCWs operating in the slow-light regime.

Such limitations may however, be overcome using hybrid slow-light PCWs interfaced with photonic nanowires [29] or PCWs interfacing between different regimes of group index [30, 31], in which the slow-light PCW is used to provide an Purcell enhancement whilst the nanowire or PCW in the fast light regime is used for photon propagation over distances >20 μm. In this way slow-light effects in waveguides may well play a leading role in developing advanced and scalable optical circuits.

3.9.1 Discussion

It has been discussed in Sect. 3.7 that the backscattering losses due to disorder in PCWs have been shown to scale as $1/v_g^2$ at high group index. From the losses observed in our PCWs up to $n_g = 36$ we estimated the localisation length for photons in PCWs when $n_g > 36$. In the region $12 < n_g \leq 36$ our observed losses had a scaling of $1/v_g$. As such, our extrapolation may have underestimated the losses at high n_g and therefore the presented values of the localisation length may represent an upper limit.

If we further consider that in practice the Purcell enhancement achieved in photonic structures can be many times less than the Purcell enhancement predicted through theoretical simulation, it is likely that group indices higher than 50 will be required to achieve a Purcell enhancement of 10, limiting further the scale of the integrated circuits to below 20 μm. Improvements in fabrication techniques should lead to an improvement in this value as disorder is reduced increasing localisation lengths and Purcell enhancements converge with the theoretical predictions.

3.10 Future Directions, Applications and Integration of Slow-light Based PCWs in Quantum Optical Circuits

Although limitations to the use of slow-light in PCWs have been identified, there are still a number of important applications for which slow-light based PCWs may be used in integrated quantum optical circuits. We have begun developing two applications which will have important consequences in the design of quantum optical circuits through the improved functionality which these applications provide.

3.10.1 PCW On-chip Phase Shifter

PCWs provide a means to control the transfer of photons within integrated photonic circuits. By predetermining an operating wavelength, the PCWs can be designed to provide required functionality, be that Purcell enhancement, high $\beta-$factors or low group velocities. However, there are a broader range of proposed applications for quantum optical circuits that require active control of the photon path where the functionality may be tuned in-situ. For example on-chip interferometry necessary for the demonstration of on-chip photon indistinguishability or in-plane polarisation transfer and readout, where the ability to control and set the phase delay of a photon path is required. This active control of slow-light on-chip also has applications in optical buffering and dynamic dispersion compensation.

It is the variation of group index with wavelength that gives a PCW the potential to be used as an on-chip phase shifter. The cut-off energy of the guided mode is sensitive to hole size and crystal geometry. Therefore by controlling the cut-off energy of the guided mode, the group index experienced by the propagating photon may be tuned. Figure 3.14 displays data for group index as a function of wavelength recorded from five 20 μm PCWs in which the r/a values vary from 0.254 to 0.274 by patterning holes with radii that vary from 65 to 70 nm whilst maintaining the

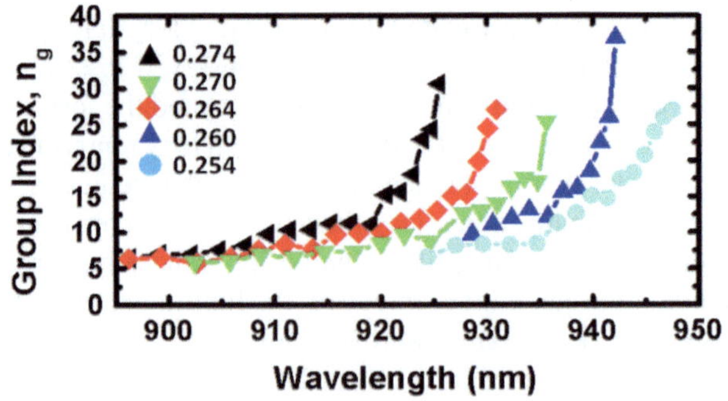

Fig. 3.14 Group index as a function of wavelength for five 20 μm long PCWs with a lattice constant $a = 255$ nm and with $r = 65$ nm (*Turquoise*), 66.25 nm (*Blue*), 67.5 nm (*Green*), 68.75 nm (*Red*) and 70 nm (*Black*). The associated r/a values are labelled

lattice constant, $a = 255$ nm. This alteration in photonic crystal geometry modifies the properties of the propagating mode such that the mode cut-off energy changes. This has the effect of shifting the $\lambda(n_g)$ curve in wavelength.

This observation leads directly to two applications. The first was described above as a solution to high losses in the slow-light mode, whereby a hybrid photonic crystal circuit is employed in which the periodicity or r/a value of the crystal is varied such that the propagating mode switches between slow-light and fast-light propagation at a fixed wavelength, as required.

The second application is active control of the photon path length by dynamic control of the photonic crystal geometry in-situ. In a simple example, temperature tuning of the PCW could provide control over the PCW geometry to act as a deterministic way of controlling the group index. The change in group index at a fixed wavelength induces a delay on a photon leaving the PCW relative to an unmodified PCW.

To demonstrate this principle, FP resonances were recorded in a PCW as a function of temperature. The temperature of the sample was externally controlled by a heating element in the cold finger cryostat. Analysis of the recorded spectra for each temperature allows the group index of the waveguide mode to be mapped.

Figure 3.15 displays the group index as a function of wavelength as extracted from the analysis of FP modes for two different r/a values and two different temperatures, 10 K (Black trace) and 50 K (red trace). It shows that there is a modification of the the group index with temperature.

A sub-section of the traces shown in Fig. 3.15 is displayed in Fig. 3.16a. It shows that at a wavelength of 929 nm, 40 K of temperature tuning provides a change in group index of 4; from $n_g = 40$ at 10 K to $n_g = 36$ at 50 K. If we consider a 1 μm long PCW this will provide a phase delay of $\sim 8\pi$ relative to an unmodified PCW.

The design of a functional device will be a compromise between a number of desired properties. For example the largest phase delays per unit of temperature tuning occur at high n_g but waveguide losses are also greatest in this region and

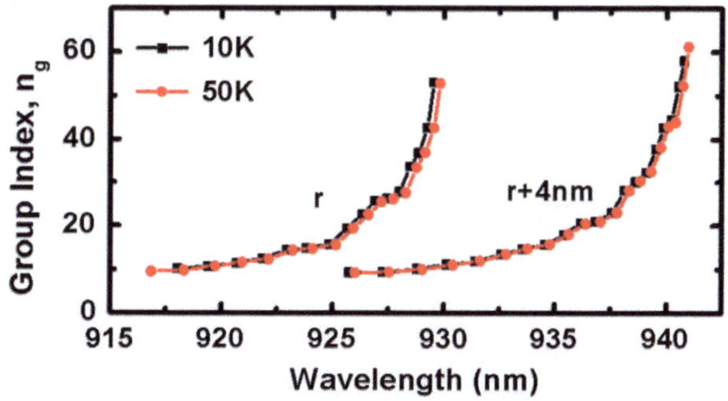

Fig. 3.15 Group Index for two PCWs recorded at temperatures of 10 K (*black*) and 50 K (*red*)

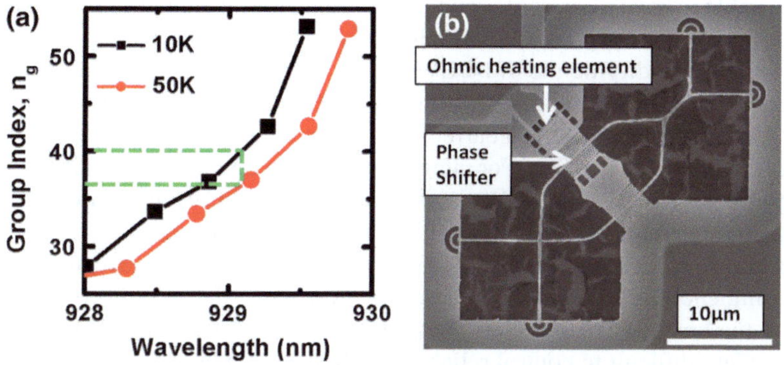

Fig. 3.16 **a** Close up of the temperature dependant group index at 929 nm from 10 K (*black*) to 50 K (*red*), displaying a change in group index of 4 at this wavelength (*green dash*). **b** SEM image of a prototype quantum optical circuit that incorporates an on-chip PCW phase shifter

working at lower group indices requires a longer PCW or larger temperature change to achieve an equivalent delay, which may limit how fast the light may become before the phase-shifters are too large to be useful.

Despite some aspects of circuit design requiring bespoke solutions to individual systems, it is likely that an in-situ heating element on-chip will be necessary to provide the active control of a temperature tuned phase-shifter. A number of photonic elements and circuits have already been designed with in-situ heating elements to control temperature on isolated elements of photonic structures, including a Mach–Zehnder interferometer in slow-light based PCWs in silicon [32].

On-chip heating elements have thus far comprised of a heating pad for laser heating [16] or an ohmic based heating element [33]. This new, but maturing, technology is part of a plan to integrate an on-chip phase shifter into a quantum optics experiment. The SEM image of a photonic circuit in Fig. 3.16b highlights the elements required to achieve the read out of a QD spin with an on-chip selective measurement basis. Although the circuit is not as yet functional, it serves as a useful tool to demonstrate how an actively controlled on-chip phase-shifter may be integrated into a quantum optics device.

3.10.2 PCW Engineering

In addition to the active control of PCW elements, the sensitivity of the waveguide mode to the PCW geometry provides an opportunity to permanently engineer desired characteristics into the waveguide mode by controlling the design of the photonic crystal.

For a number of applications in quantum optical circuits that integrate QD sources, it is desirable to have a constant group index over a large bandwidth (flat-band slow-light), as control over QD emission energy is still difficult to achieve.

The aim of modifying the W1 PCW design is to reduce dispersion and provide a constant interaction strength between the QDs and the waveguide mode over a larger wavelength window than currently exists in the PCWs studied in this chapter or typically available in PCCs. For integrated quantum optical circuits the spatial and spectral freedom offered by PCWs is an advantage over PC cavities in which spatial and spectral overlap are difficult to achieve in a single structure.

Modifications to the first two lattice periods either side of the waveguide have the largest influence over the guided modes. Modifications to the crystal geometry can be made to the hole radii or transverse/longitudinal hole positions [9]. However as the hole radii are difficult to control reliably in fabrication, adjusting the hole positions are the preferred route to achieving flat-band slow-light.

Figure 3.17 displays data from two PCWs, an unmodified structure similar to those already studied in this chapter and a modified structure in which transverse holes shifts of $0.13a$ have been incorporated into the two lattice periods of holes either side of the waveguide. In comparison to the unperturbed sample, Fig. 3.17a there is a significant modification to the the group index as a function of wavelength, Fig. 3.17b. The large variation to the simulated group index at 920 nm is due to the second guided mode crossing with the first guided mode, which is not taken into account in the simulation.

A flat-band region of moderate group index of around 30 can been seen between 935 and 945 nm.

Calculations of the Purcell factor for these flat band devices suggest that a Purcell factor (PF) $= 4$ may be achieved in a 6 nm bandwidth and even a $PF = 10$ for 1nm bandwidth, although a large group index of $\sim$70 would be required.

Coles has begun working on using longitudinal shifts of the first row of holes as it can increase the Purcell factor at a given group index. His simulations suggest that it may be possible to achieve PF $= 10$ at $n_g = 50$, which would be a considerable improvement on the PF provided by the transverse shifted PCWs [34].

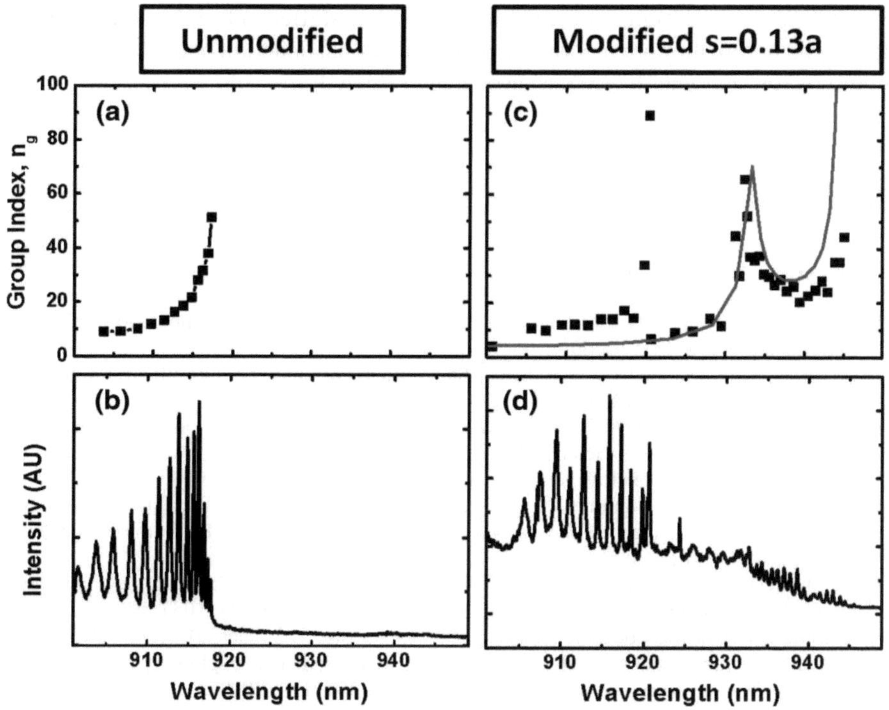

Fig. 3.17 Displays the comparison between the PL, (**b**), and group index, (**a**), extracted from an unmodified PCW and the PL (**d**) and group index (**c**) from an engineered PCW with transverse hole shifts of 0.13a. The *solid red line* in (**c**) is the expected group index extracted from simulation. Deviations form the simulation below 920 nm are due to the second guided mode crossing the guide mode of interest which is not included in the simulations

3.11 Conclusions

In conclusion, we have shown that the photon propagation lengths of GaAs QD PCWs are currently limited to a few tens of microns by unintentional disorder when operating in the slow-light regime that is capable of providing desirable Purcell enhancement. This has significant ramifications for the use of slow-light based PCWs in integrated quantum optics where photon propagation lengths many times greater may be required.

However, a number of possible applications for slow-light based PCWs in quantum optical circuits are still a possibility despite the limitations to their use. Preliminary work on two such applications, PCW phase shifters and engineered PCWs for flat-band slow-light, have been presented. The next step for our work on slow-light based PCWs is the design optimisation for the phase shifter and engineered flat-band slow-light, before they are incorporated into functional quantum optical circuits. The prototype for one such device has also been presented.

References

1. J.L. O'Brien, A. Furusawa, J. Vučković, Photonic quantum technologies. Nat. Photonics **3**, 687–695 (2009)
2. E. Knill, R. Laflamme, G.J. Milburn, A scheme for efficient quantum computation with linear optics. Nature **409**, 46–52 (2001)
3. A. Politi, J.C.F. Matthews, J.L. O'Brien, Shor's quantum factoring algorithm on a photonic chip. Science **325**, 1221 (2009)
4. I. Fushman, D. Englund, A. Faraon, N. Stoltz, P. Petroff, J. Vuckovic, Controlled phase shifts with a single quantum dot. Science **320**, 769–772 (2008)
5. T. Baba, Slow light in photonic crystals. Nat. Photonics **2**, 465–473 (2008)
6. A. Laucht, S. Pütz, T. Günthner, N. Hauke, R. Saive, S. Frédérick, M. Bichler, M.-C. Amann, A. Holleitner, M. Kaniber, J. Finley, A waveguide-coupled on-chip single-photon source. Phys. Rev. X **2**, 1–7 (2012)
7. H. Thyrrestrup, L. Sapienza, P. Lodahl, Extraction of the β-factor for single quantum dots coupled to a photonic crystal waveguide. Appl. Phys. Lett. **96**(23), 231106 (2010)
8. T. Lund-Hansen, S. Stobbe, B. Julsgaard, H. Thyrrestrup, T. Sünner, M. Kamp, A. Forchel, P. Lodahl, Experimental realization of highly efficient broadband coupling of single quantum dots to a photonic crystal waveguide. Phys. Rev. Lett. **101**, 1–4 (2008)
9. J. Li, T. White, L. O'Faolain, A. Gomez-Iglesias, Systematic design of flat band slow light in photonic crystal waveguides. Optics **16**(9), 2621–2626 (2008)
10. S. Laurent, S. Varoutsis, L. Le Gratiet, A. Lemaître, I. Sagnes, F. Raineri, A. Levenson, I. Robert-Philip, I. Abram, Indistinguishable single photons from a single-quantum dot in a two-dimensional photonic crystal cavity. Appl. Phys. Lett. **87**(16), 163107 (2005)
11. C. Santori, D. Fattal, J. Vucković, G.S. Solomon, Y. Yamamoto, Indistinguishable photons from a single-photon device. Nature **419**, 594–597 (2002)
12. A. Schwagmann, S. Kalliakos, I. Farrer, J.P. Griffiths, G.A.C. Jones, D.A. Ritchie, A.J. Shields, On-chip single photon emission from an integrated semiconductor quantum dot into a photonic crystal waveguide. Appl. Phys. Lett. **99**(26), 261108 (2011)
13. T. Ba Hoang, J. Beetz, L. Midolo, M. Skacel, M. Lermer, M. Kamp, S. HoÌĹfling, L. Balet, N. Chauvin, A. Fiore, Enhanced spontaneous emission from quantum dots in short photonic crystal waveguides. Appl. Phys. Lett. **100**(6), 061122 (2012)
14. L.O. Faolain, T.P. White, D.O. Brien, X. Yuan, M.D. Krauss, T.F. Krauss, Dependence of extrinsic loss on group velocity in photonic crystal waveguides. Opt. Express **15**(20), 13129–13138 (2007)
15. S. Hughes, J.F. Young, Extrinsic optical scattering loss in photonic crystal waveguides: Role of fabrication disorder and photon group velocity. Phys. Rev. Lett. **033903**(January), 1–4 (2005)
16. A. Faraon, I. Fushman, D. Englund, N. Stoltz, J. Vučković, Dipole induced transparency in waveguide coupled photonic crystal cavities. Opt. Express **16**(16), 944–947 (2008)
17. E. Kuramochi, M. Notomi, S. Hughes, A. Shinya, T. Watanabe, L. Ramunno, Disorder-induced scattering loss of line-defect waveguides in photonic crystal slabs. Phys. Rev. B **72**, 2–5 (2005)
18. J. Topolancik, B. Ilic, F. Vollmer, Experimental observation of strong photon localization in disordered photonic crystal waveguides. Phys. Rev. Lett. **99**, 2–5 (2007)
19. L. Sapienza, H. Thyrrestrup, S. Stobbe, P.D. Garcia, S. Smolka, P. Lodahl, Cavity quantum electrodynamics with Anderson-localized modes. Science **327**, 1352–1355 (2010)
20. S. Johnson, J. Joannopoulos, Block-iterative frequency-domain methods for Maxwell's equations in a planewave basis. Opt. Express **8**, 173 (2001)

21. P. Anderson, Absence of diffusion in certain random lattices. Phys. Rev. **386**(1949), 1492–1505 (1958)
22. H. Thyrrestrup, S. Smolka, L. Sapienza, P. Lodahl, Statistical theory of a quantum emitter strongly coupled to Anderson-localized modes. Phys. Rev. Lett. **108**, 1–5 (2012)
23. P. García, S. Smolka, S. Stobbe, P. Lodahl, Density of states controls Anderson localization in disordered photonic crystal waveguides. Phys. Rev. B **82**, 1–5 (2010)
24. C.W.J. Beenakker, Random-matrix theory of quantum transport. Rev. Mod. Phys. **69**(3), 731–808 (1997)
25. A. García-Martín, J. Sáenz, Universal conductance distributions in the crossover between diffusive and localization regimes. Phys. Rev. Lett. **87**, 116603 (2001)
26. R. Patel, A. Bennett, K. Cooper, P. Atkinson, C. Nicoll, D. Ritchie, A. Shields, Postselective two-photon interference from a continuous nonclassical stream of photons emitted by a quantum dot. Phys. Rev. Lett. **100**, 1–4 (2008)
27. E.B. Flagg, A. Muller, S.V. Polyakov, A. Ling, A. Migdall, G.S. Solomon, Interference of single photons from two separate semiconductor quantum dots. Phys. Rev. Lett. **104**, 1–4 (2010)
28. V. Manga Rao, S. Hughes, Single quantum-dot Purcell factor and β factor in a photonic crystal waveguide. Phys. Rev. B **75**, 1–7 (2007)
29. J.P. Hugonin, P. Lalanne, T.P. White, T.F. Krauss, Coupling into slow-mode photonic crystal waveguides. Opt. Lett. **32**, 2638–2640 (2007)
30. T.P. White, L.C. Botten, C. Martijn de Sterke, K.B. Dossou, R.C. McPhedran, Efficient slow-light coupling in a photonic crystal waveguide without transition region. Opt. Lett. **33**, 2644–2646 (2008)
31. S. Mazoyer, P. Lalanne, J. Rodier, J. Hugonin, M. Spasenović, L. Kuipers, D. Beggs, T. Krauss, Statistical fluctuations of transmission in slow light photonic-crystal waveguides. Opt. Express **18**, 14654 (2010)
32. Y.A. Vlasov, M. O'Boyle, H.F. Hamann, S.J. McNab, Active control of slow light on a chip with photonic crystal waveguides. Nature **438**, 65–69 (2005)
33. A. Faraon, J. Vučković, Local temperature control of photonic crystal devices via micron-scale electrical heaters. Appl. Phys. Lett. **95**(4), 043102 (2009)
34. J. Liang, L.-Y. Ren, M.-J. Yun, X. Han, X.-J. Wang, Wideband ultraflat slow light with large group index in a W1 photonic crystal waveguide. J. Appl. Phys. **110**(6), 063103 (2011)

Chapter 4
On-chip Interface for In-plane Polarisation Transfer for Quantum Information Processing

4.1 Introduction

A network of quantum dot (QD) spins connected via photons is a promising system for optical quantum information processing [1].

Conceptually, the simplest method of communicating the spin state of a QD is to use the polarisation of a photon emitted along the growth direction, with circular polarisation states mapping to the spin up/down states. In practice, using out-of-plane photons requires external optics which becomes unwieldy as the number of nodes increases.

In order to implement this in a scalable architecture, a lithographically defined waveguide circuit is required. However, the spin degree of freedom is optically accessed via the photon polarisation and because the QD dipole lies in the plane a single planar waveguide cannot transmit circular polarisation. The single in-plane waveguide cannot support the longitudinal polarisation component and therefore polarisation information is lost.

Here we introduce and demonstrate a crossed waveguide device, where the polarisation of a photon emitted by a QD at the intersection is converted to a 'which path' state, with the x(y)-polarisation component transmitted along the y(x)-direction waveguide. By recombining the x- and y-direction waveguides the polarisation of the photon can be recovered and the spin state of the QD deduced.

A scalable optical quantum information processor [2–4] is likely to be a waveguide circuit [5] with integrated sources [6–8] detectors [9] and either deterministic quantum memory logic or quantum memory elements [3, 10]. This work provides a promising way forward for interfacing solid state spins with optical waveguide circuits.

Photons are the most robust carriers of quantum information and the most easily manipulated at the single qubit level [2]. With the recent demonstration of Shor's algorithm using several one- and two-qubit gates in a single waveguide circuit [4], linear optical quantum computing is an early front runner. In such a linear optics approach, measurements are used to implement two-photon gates resulting in a probabilistic

N. A. Wasley, *Nano-photonics in III–V Semiconductors for Integrated Quantum Optical Circuits*, Springer Theses, DOI: 10.1007/978-3-319-01514-9_4,

operation [3]. However, without a deterministic single photon nonlinearity [11] or a quantum memory such an approach is intrinsically un-scalable [3, 10]. This can be overcome with a solid-state emitter, for example QDs, nitrogen-vacancy centres in diamond or impurity centres in semiconductors, which can be integrated within the waveguide circuit.

III-V semiconductor QDs are promising solid-state quantum emitters, with strong optical dipole, and lifetime limited radiative recombination [12]. Highly tuneable QDs [13, 14] are easily integrated with photonic structures: non-classical light sources [6–8], strong-coupling [15, 16], indistinguishable photons from two remote QDs [17, 18] and on-chip integration with single photon detectors [9] have all been demonstrated.

The QD states most suitable as storage qubits are the spin eigenstates of an electron or hole, with intrinsic coherence times in the microsecond regime [19, 20]. Optical detection and manipulation [21, 22] of the spin is achieved by mapping to the circular polarisation state of a photon, i.e. $(\alpha\ |\uparrow\rangle + \beta e^{i\phi}\ |\downarrow\rangle) \Rightarrow (\alpha\ |\ \sigma^-\rangle + \beta e^{i\phi}\ |\ \sigma^+\rangle)$. However, interfacing the spin $(\uparrow, \downarrow)$ to a planar waveguide is problematic as only the x or y component of the left (σ^-) and right (σ^+) circularly polarised light will propagate.

In this chapter, a device for transferring a QD spin state to an in-plane photon state is presented. To demonstrate its operation the spin of a QD is deduced from photons emitted in the sample plane.

4.2 Principle of the Interference Experiment to Demonstrate Polarisation Detection

The principle of the experiment used to detect the polarisation is described here using a schematic diagram of a simple optical set-up, consisting of a Mach Zehnder Interferometer (MZI) comprised of one polarising beam splitter (PBS) and one non-polarising beam splitter (NPBS), shown in Fig. 4.1a.

Circularly polarised light incident on the PBS is split into x and y-polarised components which propagate in the two arms of the MZI. For $\sigma^+(\sigma^-)$ light the y-polarised component has a relative phase to the x-polarised component of $\pi/2(-\pi/2)$. The y-polarised light is rotated by the half wave-plate $(\lambda/2)$ so both components have parallel polarisation and then the light from the two arms of the MZI is interfered on the NPBS and the signal detected on two detectors D1 and D2.

As the path delay between light in the two arms is modulated by the piezo-actuated mirror, interference fringes are recorded on the two detectors, D1 and D2, as illustrated by Fig. 4.1b.

Any unequal phase shifts that occur in the arms of the interferometer due to reflections at mirror surfaces or unequal path lengths within optical fibres (if they were to be used) affect the $\sigma^+(\sigma^-)$ polarised components in that arm equally, and as such any effective phase change can be incorporated into the path delay and it therefore

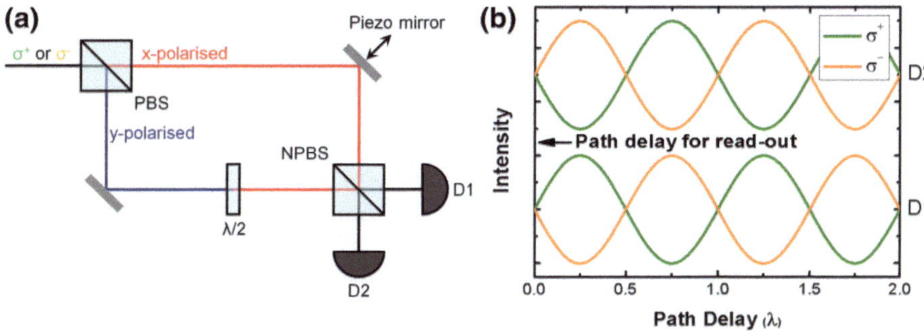

Fig. 4.1 Shows the principle of operation of the polarisation to which path interface. **a** Mach-Zehnder Interferometer comprised of one polarising beam splitter (PBS), one non-polarising beam splitter (NPBS), half wave plate ($\lambda/2$) and two detectors (D1 and D2). **b** Interference fringes recorded on D1 and D2 as the path delay in one arm of the MZI is modulated using a piezo controlled mirror

does not affect the relative phase of the interference fringes. However, the π phase shift that occurs on reflection at the NPBS results in a π phase difference between the interferences fringes recorded on D1 and D2. Therefore, the first interference maxima (minima) for $\sigma^+(\sigma^-)$ light occurs at a phase delay of $\pi/2$ which is a path delay of $\lambda/4$ as displayed in Fig. 4.1b. As a result, when the path delay is set at $\lambda/4$, σ^+ light incident on the MZI can be detected by a high signal level on D1 and σ^- by a high signal level on D2.

4.2.1 Experimental Demonstration Using a QD in the Bulk

To demonstrate the principle of the experiment and to explain the method used to analyse the data, light emitted vertically by a single QD located in an unprocessed area of the wafer and an experimental set-up constructed from table-top optics is presented first, prior to the in-plane transfer which is to follow.

In this demonstration PL emitted from Zeeman split energy levels of a single QD that are circularly polarised (σ^+ / σ^-) as labelled in Fig. 4.2a is fed into the MZI pictured in Fig. 4.2b. The fiber beam-splitter (FBS) ensures a good overlap of the optical modes. The light output from one port of the FBS is spectrally filtered using a 0.55 m single grating spectrometer and directed onto a pair of silicon avalanche photo-diodes (APDs); one for each Zeeman split component. The spectral window detected by each APD is colour coded orange (σ^-) and green (σ^+) in Fig. 4.2, this two colour selection is achieved as described in Chap. 2, Sect. 2.7.4.

The measurement is performed by varying the delay in one arm in steps of ~100 μm using the delay stage labelled in Fig. 4.2b. At each point the time delay is varied over a few wavelengths using the piezo-actuated mirror and the self interference of the two transitions are recorded on the separate APDs. Figure 4.3a shows a

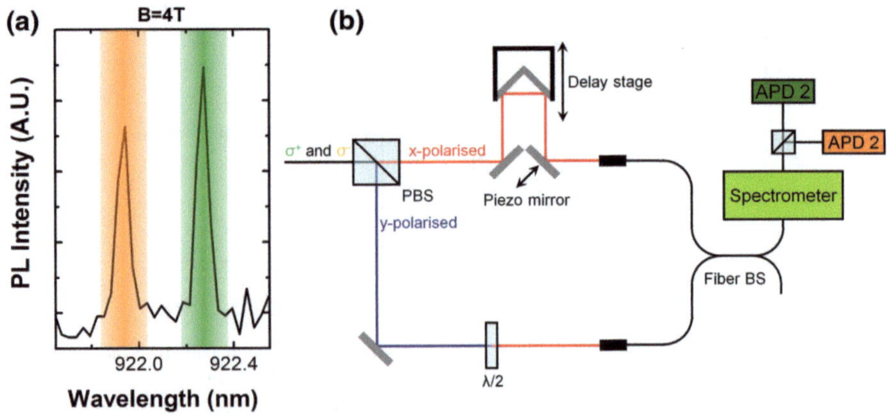

Fig. 4.2 **a** μPL spectrum showing the Zeeman split transitions of a single QD with an applied magnetic field, $B = 4T$. The *shaded regions*, *orange* and *green* indicate the spectral window collected on the two colour coded APDs, *orange* (σ^-) and *green* (σ^+). **b** Experimental set-up used to demonstrate polarisation to which path interface using bulk optics

typical plot of the intensity recorded on the two APDs as the path delay is varied over a few wavelengths.

It can be seen from Fig. 4.3a that the period of the oscillations is not constant, which results from mechanical fluctuations in the experimental set-up.

This instability in the set-up is overcome by plotting the intensity of the σ^- transition as a function of σ^+, as shown in Fig. 4.3b. This method circumvents the need for a stabilised interferometer as it is the relative phase, $\Delta\phi$, between the interference fringes of the σ^+ and σ^- transitions, which is of interest.

The Zeeman splitting means that as the delay time of the interferometer is changed, $\Delta\phi$ varies between zero and π, with a period proportional to the Zeeman splitting. The data presented in Fig. 4.3a is taken at a delay time when the two transitions are approximately in phase.

This variation in phase causes a variation in the shape of the $\sigma^\pm$ intensity plot. The shape evolves from a straight line with a negative gradient when $\Delta\phi = \pi$, to a straight line with a positive gradient when $\Delta\phi = 0$, via a circle.

When in phase, $\Delta\phi = 0$, a high (low) signal on the APD recording the σ^+ interference is matched with a high (low) signal on the APD recording the σ^- interference, creating a straight line with a positive gradient on the $\sigma^\pm$ intensity plot, as shown in Fig. 4.3a, b.

When out of phase, $\Delta\phi = \pi$, a high (low) signal on the APD recording the σ^+ interference is contrasted with a low (high) signal on the APD recording the σ^- interference, creating a straight line with a negative gradient on the $\sigma^\pm$ intensity plot, as shown in Fig. 4.3c, d.

Figure 4.4a–e shows intensity plots for five different time delays and demonstrates the expected variation in the shape of the plot as $\Delta\phi$ evolves.

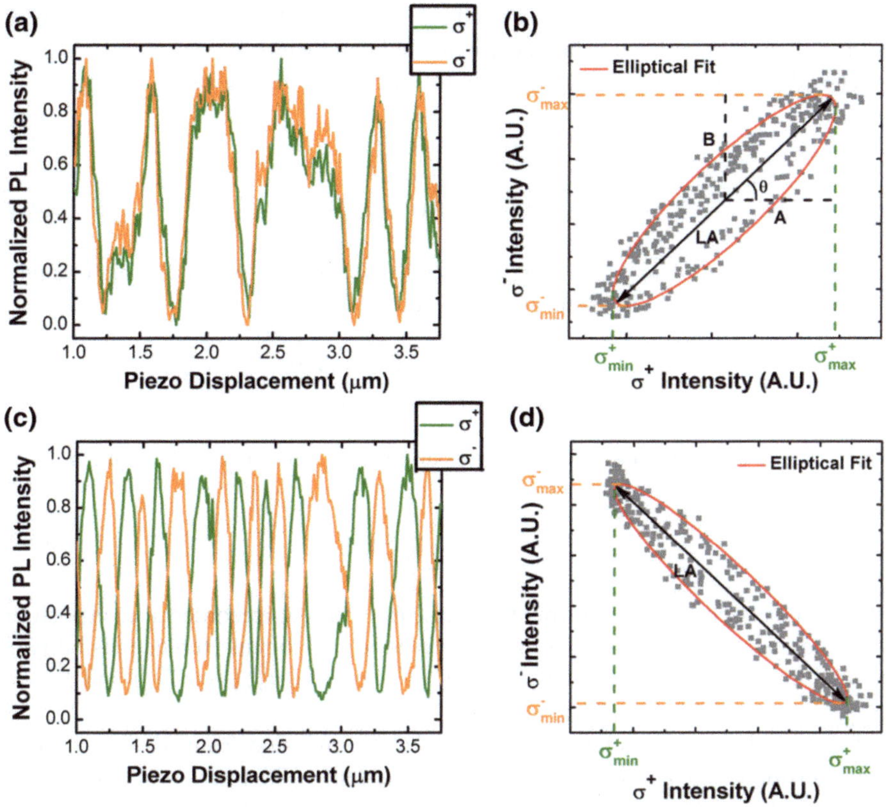

Fig. 4.3 **a** Interference fringes recorded when $\Delta\phi = 0$ and the σ^+ and σ^- polarised transitions are in phase, as the path delay in one arm of the interferometer is varied using the piezo-actuated mirror. **b** Scatter plot showing intensity of σ^- transition plotted against intensity of σ^+ transition for data in **a**. The *red line* shows an elliptical fit to the data, and the values LA, B, A and θ are defined for use in extracting $\Delta\phi$. **c**, **d** Display the corresponding plots to **a**, **b** for the case where $\Delta\phi = \pi$ and the σ^+ and σ^- polarised transitions are out of phase

Extraction of Relative Phase from Interference Experiments

To extract the relative phase, $\Delta\phi$, and the interference visibility the data is fit using the conic equation of the ellipse:

$$ax^2 + bxy + cy^2 + dx + ey + f = 0 \tag{4.1}$$

where a, b, c, d, e and f are fitting parameters determined by the shape, size and orientation of the ellipse.

From this fit, the following parameters characterising the ellipse can be extracted, as defined in Fig. 4.3b. The tilt angle, θ, is the angle between the x-axis and the long axis of the ellipse, LA. The extreme maximum (minimum) values of the ellipse for the σ^+ and σ^- transitions are designated $\sigma^+_{max}(\sigma^+_{min})$ and $\sigma^-_{max}(\sigma^-_{min})$, respectively. $\Delta\phi$ can then be calculated using these extracted parameters.

The ellipse can also be defined by the following two equations,

$$x(t) = A\cos(t) \tag{4.2}$$

$$y(t) = B\cos(t + \Delta\phi) \tag{4.3}$$

where $0 \leq t < 2\pi$, $A = (\sigma^+_{max} - \sigma^+_{min})/2$ and $B = (\sigma^-_{max} - \sigma^-_{min})/2$. The tilt angle, θ can also be expressed in terms of x and y at $t = 0$,

$$\tan(\theta) = \frac{y(0)}{x(0)} = \frac{B\cos(\Delta\phi)}{A} \tag{4.4}$$

which can be rearranged to give an expression for $\Delta\phi$ in terms of θ, A and B,

$$\Delta\phi = \cos^{-1}\left[\frac{A}{B}\tan(\theta)\right] \tag{4.5}$$

which can be used to extract the relative phase from the data recorded at each position of the delay stage. The elliptical fit is also used to find the interference visibility of the two transitions, defined as $V_{\sigma^+} = (\sigma^+_{max} - \sigma^+_{min})/(\sigma^+_{max} + \sigma^+_{min})$ and $V_{\sigma^-} = (\sigma^-_{max} - \sigma^-_{min})/(\sigma^-_{max} + \sigma^-_{min})$.

Figure 4.4f plots the relative phase, $\Delta\phi$ of the emission from the two Zeeman split transitions against the interferometer time delay. The interference of the two transitions move in and out of phase as the delay in the MZI is varied. The period of this beating is equal to 8.4 ps, corresponding to the Zeeman energy-splitting of 0.48 meV at $B = 4T$. At an interferometer time delay of zero, $t_d = 0$, $\Delta\phi = 0.95\pi$, very close to the expected value of π.

To ensure the zero delay time is known to a high accuracy an interference measurement is performed using light from the QD distribution, excited at high power and spectrally filtered using only a 900 nm long pass filter. The QD distribution is incoherent so the interference visibility decays rapidly as the stage is moved away from $t_d = 0$. Figure 4.4f also shows the results from this measurement and a fit to the data of the function $V(t) = V_0 + Ae^{\frac{(t-t_0)}{\tau}}$, which identifies the position of the delay stage for $t_d = 0$ to an accuracy of $\pm 20\,fs$ ($\pm$0.003 mm), determined by the repeatability of the stage.

The coherence of the QD can also be determined by measuring the interference visibility to longer time delays using Fourier transform spectroscopy. Figure 4.4g shows a plot of the interference visibility against delay time. The solid red line shows a fit to the data [23], from which the linewidth of the QD can be extracted. Values of 12.8 μeV and 12.7 μeV (~150 ps) are found from the σ^+ and σ^- transitions, respectively.

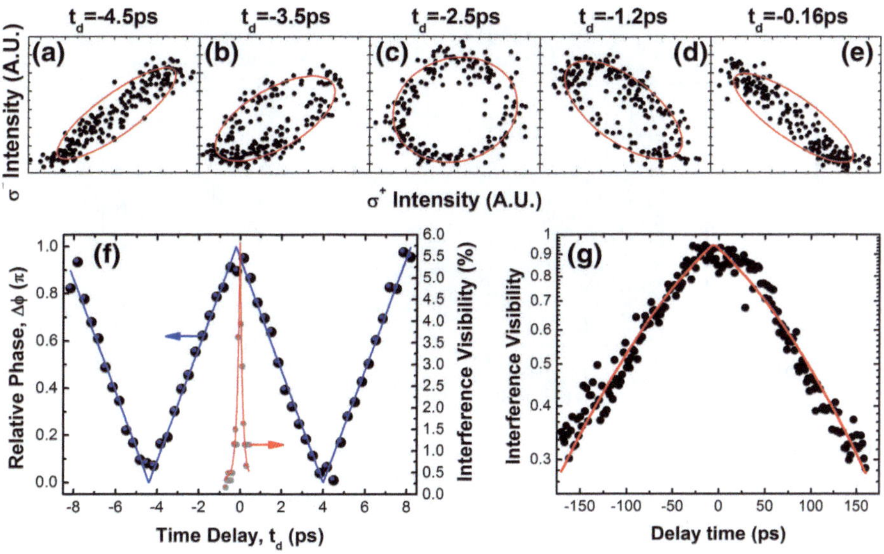

Fig. 4.4 **a–e** Intensity plots of *right* and *left* circularly polarised states interfered simultaneously on two detectors. **f** Plot of relative phase between the two transitions as the path delay in one arm of the MZI is varied. The *line* shows a triangular waveform fit to the data. **g** Coherence measurement of the σ^+ transition using Fourier transform spectroscopy, a linewdith of 12.8 μeV (~150 ps) is extracted

4.3 Sample Details

Having demonstrated the principle of the experiment in the bulk, the prototype device for in-plane polarisation transfer may now be introduced. The samples used in this study are grown by molecular beam epitaxy (MBE) on undoped GaAs (100) wafers. The wafer consists of a 140 nm GaAs waveguide slab with a single layer of nominally InAs QDs at its centre, separated from the GaAs buffer layer by a 1 μm thick $Al_{0.6}Ga_{0.4}As$ sacrificial layer. A rotation stop is included during the InAs deposition in order to achieve a low QD density. This results in a grading of the QD density across the sample and a minimum measured density of $\sim 1 \times 10^9\,cm^{-2}$.

This area of the sample is then employed to fabricate two orthogonal free-standing waveguides with a width of ~200 nm connected to four out-couplers. Electron beam lithography is used to define the waveguide devices and the GaAs slab layer etched using a chlorine based reactive ion etch (RIE). Finally, hydrofluoric acid is used to selectively remove the $Al_{0.6}Ga_{0.4}As$ layer from beneath the waveguides leaving the free-standing waveguide structure.

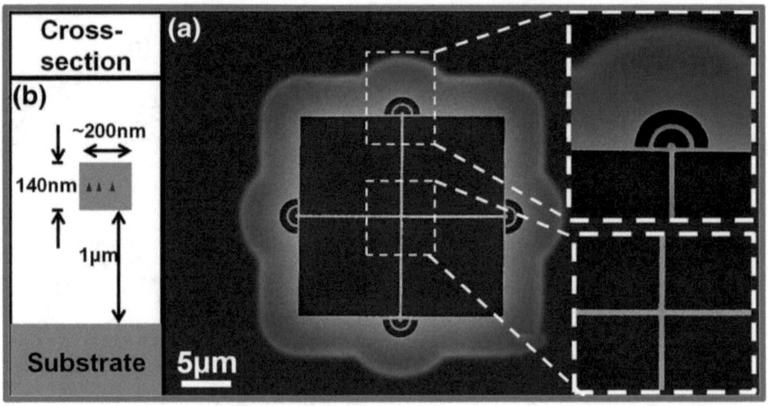

Fig. 4.5 **a** SEM image of a fabricated device. The crossed waveguides are terminated by four out-couplers. Inset, selected zoomed sections to highlight fabrication quality. **b** Schematic cross section of the prototype device. Free standing waveguides (nanowires) 200 nm wide and 140 nm deep contain a single layer of InAs QDs

4.4 Prototype Crossed Waveguide Device

Figure 4.5a shows a scanning electron microscope (SEM) image of a prototype spin-waveguide interface. This device consists of two orthogonal free-standing waveguides (nanowires) with a width of ~200 nm connected to four out-couplers [24]. The out-couplers are employed to scatter light that propagates along the waveguide into the vertical direction for collection. The waveguides are fabricated from a 140 nm thick GaAs layer containing a single layer of InGaAs QDs at its centre as displayed schematically in Fig. 4.5b.

A QD located at the centre of the waveguide intersection will coherently emit the x(y)-polarisation component of a circularly-polarised state into the waveguides aligned along the y(x)-directions respectively. By collecting both polarisation components, whilst retaining their relative phase, the full polarisation state of the photon is mapped to a which-path state. On recombining the light from the waveguides, the polarisation state of the photon can be reconstructed at another point in the plane, hence enabling on-chip transfer of spin information.

4.5 Optical Measurements

The optical measurements, shown schematically in Fig. 4.6, are performed in a low temperature confocal microscope system equipped with a superconducting magnet described in detail in Chap. 2. In brief, the photoluminescence is excited using an 855 nm continuous wave (CW) diode laser, linearly polarised at 45° to the x-axis, as defined in Fig. 4.6 and focused to a ~1 μm diameter.

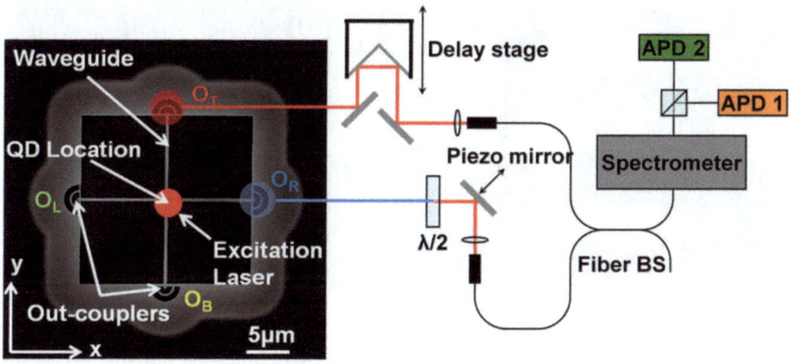

Fig. 4.6 Schematic diagram of the optical set-up used for the self interference of the two Zeeman split QD transitions. Spatially selective collection spots positioned in the sample plane collect PL from orthogonal out-couplers O_T and O_R. PL is excited by a laser at the positioned at the intersection. The two optical collection paths are recombined by a single mode fiber beam-splitter, ensuring good spatial overlap of the modes. Interference fringes are recorded for each QD transition on separate APDs having been isolated in energy by the spectrometer. The path length is modulated by a piezo-actuated mirror and motorised delay stage

The optical system, pictured in Fig. 4.6 allows us to spatially select which region of the sample the emission is collected from and independently control the position of the excitation laser spot.

The spatial selection in the collection path is achieved with a pair of single mode optical fibers which can be independently positioned in the image-plane to simultaneously collect emission from two discrete locations on the sample as described in Chap. 2, Sect. 2.7.1. For the interference experiments the light from the two orthogonal out-couplers, O_R and O_T labelled in Fig. 4.6, is coupled into a polarisation maintaining single mode fiber beam-splitter. A half waveplate is set in one arm of the interferometer to ensure maximum interference at the beam-splitter. The light from one output port of the fiber beam-splitter is filtered with a 0.55 m spectrometer and monitored using a pair of silicon avalanche photo-diodes (APDs), which can be individually tuned to detect two different wavelengths simultaneously as detailed in Chap. 2, Sect. 2.7.4. The path delay in the interferometer is modulated using a linear delay stage and piezo-actuated mirror.

It can be noted that the optics are similar to the optics displayed schematically in Fig. 4.2 with the exception that the first PBS in Fig. 4.2 is in effect replaced by the crossed waveguide structure.

4.6 Device Characterisation

To characterise the function of the device one of the collection fibers is aligned to the excitation laser (that is near resonant with the QD wetting layer, 855 nm) so that the collection and excitation are occurring at the same location on the sample.

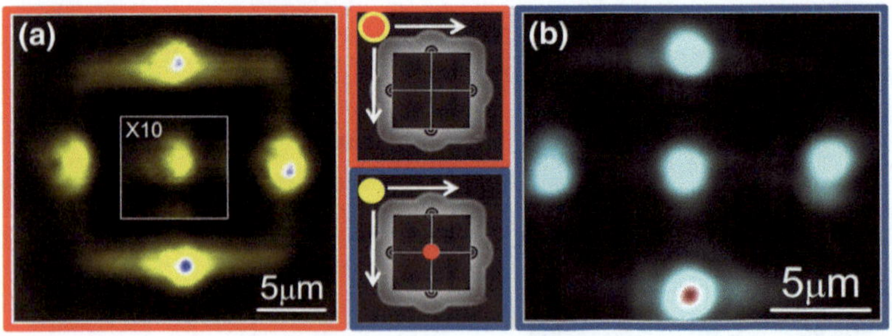

Fig. 4.7 PL intensity maps, integrated over the QD distribution, recorded from the crossed waveguide device, by **a** scanning the collection fiber and excitation laser together across the structure as shown schematically in the inset (*red border*) and **b** scanning the collection fiber with the excitation laser fixed at the intersection as shown in the inset (*blue border*)

The objective lens is scanned across the sample whilst the PL emission, filtered by a 900 nm long pass filter, is recorded on an APD. This allows a map of the QD emission intensity from the sample to be made.

Figure 4.7a shows a PL map in which the brightest emission is observed from the four out-couplers. The emission intensity from the central region, shown in the white box, is multiplied by a factor of 10 for clarity. This demonstrates the effectiveness of the out-couplers in scattering the QD PL vertically. The emission from the out-couplers is ~25X brighter than from the central region of the crossed waveguides and 100X brighter than the un-processed wafer. This supports recent reports which also show greatly improved vertical collection efficiency resulting from circular gratings [25] and highlights the suitability of the out-couplers for coupling QD emission into a single mode fiber.

Mapping the structure in this way also allows us to locate the position of the centre of the intersection, which is necessary for all subsequent measurements.

With the laser positioned at the intersection the position of the collection optics is then scanned. This allows us to observe the emission from the structure when exciting at the intersection of the waveguides. Figure 4.7b shows a PL map taken by scanning the collection spot in this way, and shows strong emission from the four out-couplers and the point of excitation. This demonstrates that the QD emission excited at the central region is transmitted along the waveguides and scattered vertically by the out-couplers.

Under these conditions the intensity of the PL emission collected from the out-couplers is approximately equivalent to the PL intensity collected from the position of excitation.

The polar plot of the emission from the structure in Fig. 4.8 shows that the emission from the out-couplers is, as expected, strongly polarised perpendicular to the direction of propagation of the waveguide. The emission from the centre is unpolarised, as it couples equally to the O_T and O_R waveguides. This polarisation of the light from the out-couplers helps to simplify the collection optics as it allows us to separate the light from O_T and O_R using a polarising beam-splitter (PBS) in the collection path.

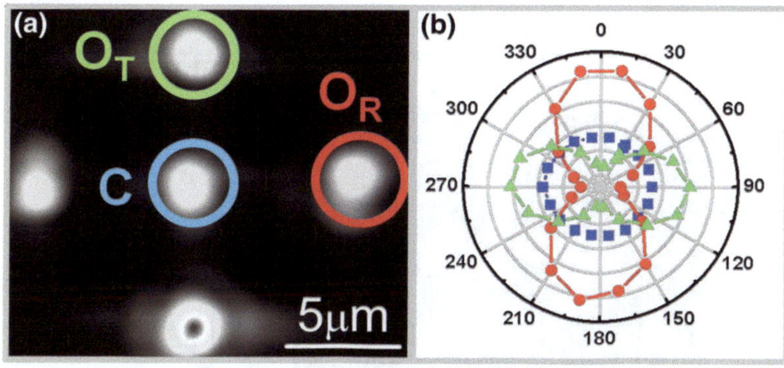

Fig. 4.8 **a** The PL map from Fig. 4.7b is used here to mark the three areas of strong emission: the crossed waveguide intersection (C (*blue*)) and two orthogonal out-couplers (O_T(*green*) and O_R(*red*)), from which polarisation dependent collection measurements are taken. **b** The two out couplers are strongly linearly polarised, orthogonal to one another, whilst PL from intersection is unpolarised

It is important to note that the only role of the PBS is to simplify the spatial selectivity of the system and could equally be replaced with a mirror in these experiments.

4.7 Demonstration of In-plane Polarisation Transfer

4.7.1 Identification of Suitable QD Transitions

Once the device has been fully characterised, the next stage is to identify a device in which a single QD emits into two orthogonal waveguides.

This is achieved by exciting the structure at the intersection and recoding PL spectra from the intersection and the out-couplers, labelled C, O_T and O_R in Fig. 4.8. The spectra are then compared so that a single QD transition that is visible from each collection point may be identified. This indicates the QD is located at the intersection and is coupled to each waveguide.

4.7.2 Hanbury-Brown and Twiss Autocorrelation Measurement

To verify that the QD transitions that are studied originate from a single QD a photon correlation measurement using a Hanbury-Brown and Twiss (HBT) set-up was performed on the QD transition prior to the application of a magnetic field. By demonstrating the initial transition is a single photon emitter we can conclude that the subsequent Zeeman split transitions originate from one QD.

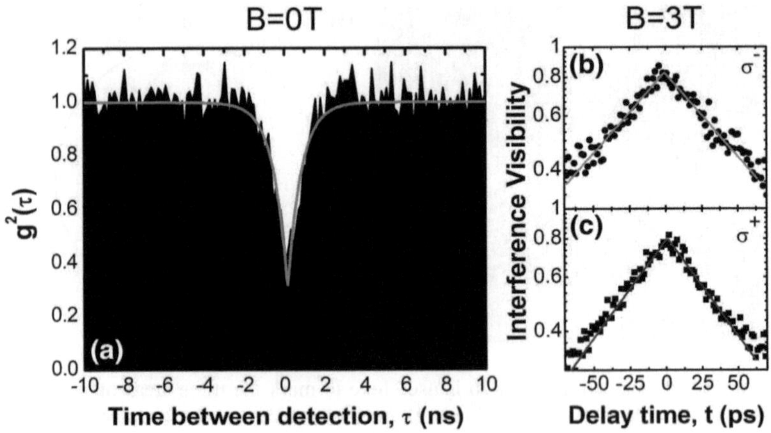

Fig. 4.9 **a** Photon correlation histogram of the QD transition at $B = 0T$ using a Hanbury-Brown and Twiss set-up. The *red-line* shows a fit to the data of the function $g^{(2)}(\tau) = (1-(1-M)e^{(-|\tau|/\tau_d)})$, where $M = 0.28$ is the multi-photon emission probability and $\tau_d = 0.76$ ns is the lifetime of the QD. Coherence time measurements of the σ^- (**b**), and σ^+ (**c**), transitions respectively. The *solid red* and *blue lines* show fits to the data following Ref. [23]

The QD emission is collected from the intersection of the waveguides and filtered using a spectrometer, before being split with a beam-splitter and coincidence counts recorded using a pair of APDs.

The histogram recorded of the relative time delay, τ between a photon detection at one APD and the other is shown in Fig. 4.9a. The red-line shows a fit to the data of the function $g^{(2)}(\tau) = (1-(1-M)e^{(-|\tau|/\tau_d)})$. The measurement shows clear anti-bunching with a multi-photon emission probability, $g^{(2)}(0) = 0.28$ obtained with no background subtraction. Verifying that the QD transition originates from a single QD.

The lifetime of the QD transition is also evaluated from this measurement, $\tau_d = 0.76$ ns.

4.7.3 Application of Magnetic Field

A magnetic field, $B = 3T$ is then applied normal to the sample plane (Faraday geometry), so that the σ^+ and σ^- polarised transitions are separated by the Zeeman energy splitting and can be identified by their characteristic energies.

Figure 4.10a shows the μPL spectrum for a single QD transition that is observable from the Centre, O_T and O_R as an increasing magnetic field is applied. The Zeeman split doublet originating from a QD located at the intersection can be observed from all four out-couplers. The spectra recorded at $B = 3T$ from O_T and O_R are displayed in Fig. 4.10c and d.

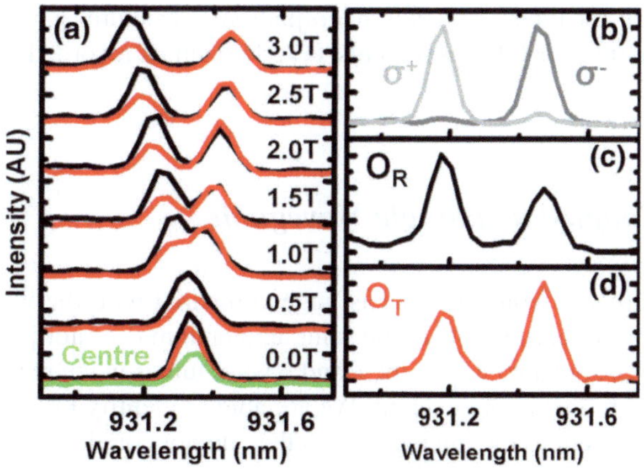

Fig. 4.10 **a** μPL spectra of a single quantum dot transition recorded from the crossed waveguide intersection, *Centre* (*green*), and two orthogonal out-couplers O_T(*red*) and O_R (*black*). Spectra are also displayed of the Zeeman split QD transition observed from O_T and O_R as a magnetic field is applied normal to the plane of the sample. **b** Spectra of the Zeeman split QD transition at $B = 3T$ observed from the centre with polarisation sensitive detection of σ^+ (*light grey*) and σ^- (*dark grey*). **c** and **d** are the spectra of the Zeeman split QD transition at $B = 3T$ recorded from O_R and O_T respectively

When observed vertically from the intersection, polarisation sensitive detection confirms that the two transitions are right and left circularly polarised, shown in Fig. 4.10b.

For this demonstration the spin-state of the QD refers to the exciton pseudo-spin states of the electron and hole spin ($\uparrow, \Downarrow$) and ($\downarrow, \Uparrow$) which also map to circularly polarised photon states.

The contrast, C_α, between the $\sigma^\pm$ transitions observed from the out-couplers, defined as $C_\alpha = (I^+ - I^-)/I^+ + I^-)$, where $I^+(I^-)$ is the PL intensity of the $\sigma^+(\sigma^-)$ transition and $\alpha = T, R$ are $C_R = -0.19$, and $C_T = +0.19$ respectively. This is consistent with mapping horizontal and vertical linear polarisations to the orthogonal waveguides, and is close to the desired value of $C_\alpha = 0$ which is the case when both transition couple equally to both waveguides.

4.7.4 Fourier Spectroscopy: QD Coherence Time

The coherence length of the QDs can be extracted from the interference measurements, as previously discussed using Fourier transform spectroscopy.

Figure 4.9b and c plot the decay of the interference visibility against the time delay for the $\sigma^+$and σ^- transitions. The data is fitted following Ref. [23] and coherence times of 81 and 68 ps (16.2 and 19.4 μeV) are extracted for the $\sigma^+$and σ^- transitions, respectively.

This is typical for InGaAs QDs in nano-photonic structures [26, 27] but shorter than values reported for QDs in the bulk [17, 18] and as recorded for a QD in the bulk in Sect. 4.2.1.

4.7.5 QDs Coupled to a Single Waveguide

As the excitation spot from the laser is large compared with the waveguide width and the density of quantum dots in the sample, quantum dots not just in the central region of the cross, but also within the x and y waveguides in the regions adjacent to the intersection are inevitably excited. We are able to identify QDs that are located in the regions adjacent to the intersection as they do not couple to both waveguides. Figure 4.11a shows a situation where a QD in the x waveguide is excited and PL spectra from O_T, O_R, and the excitation region *Centre* is observed. The QD at 1.3623 eV is clearly observed from O_R and the *Centre*, but not from O_T.

The light from O_R and the *Centre* are cross-polarised (Fig. 4.11b) suggesting that y-polarised emission from the QD is coupled to the waveguide and observed from the out-coupler, whereas the x-polarised light cannot propagate in this direction so is emitted vertically.

4.7.6 In-plane Deduction of Photon Polarisation

Finally, to demonstrate that the full polarisation of the emitted photon is coherently mapped to the which-path state, we show that the light emitted from two orthogonal out-couplers is mutually coherent. The interference experiment described in

Fig. 4.11 **a** μPL spectra recorded from the crossed waveguide intersection, *Centre* (*black*), and two orthogonal out-couplers O_T (*red*) and O_R (*blue*). **b** Intensity of the QD transition shown in **a** with polarisation dependent detection from the *Centre* and O_R. The transition is cross polarised between the *Centre* and O_R whilst not being visible from O_T. This is consistent with a QD coupled to only one waveguide when located off centre and not in the intersection

Sect. 4.2.1, is performed using two fibers positioned to collect emission from O_R and O_T.

The x and y-polarised components of the photons emitted by the QD are each split by the waveguide intersection and propagate down the waveguides (the crossed waveguides act as the first beam splitter in the interferometer).

The two paths are recombined at a fiber beam-splitter, as depicted in Fig. 4.6b, forming an interferometer. By varying the path difference by a few wavelengths the interference fringes are recorded simultaneously for the σ^+ and σ^- transitions on separate APDs.

An example of this data, where the $\sigma^\pm$ signals are in anti-phase is presented in Fig. 4.12a.

To deduce the relative-phase, $\Delta\phi$ between the $\sigma^\pm$ lines for each coarse time-delay t_d, a phase-plot of I^+ against I^- is constructed (Fig. 4.12b). The elliptical fit to the data allows $\Delta\phi$ to be extracted using the analysis given in Sect. 4.2.1.

Figure 4.12d displays phase-plots at five different time delays. As the delay is varied from −0.8 to 4.2 ps, the ellipticity of the plot evolves from a straight line with negative gradient ($\Delta\phi = \pi$) to a straight line with positive gradient ($\Delta\phi = 0$), via a circle ($\Delta\phi = \pi/2$), as a result of the Zeeman splitting between the $\sigma^\pm$ lines.

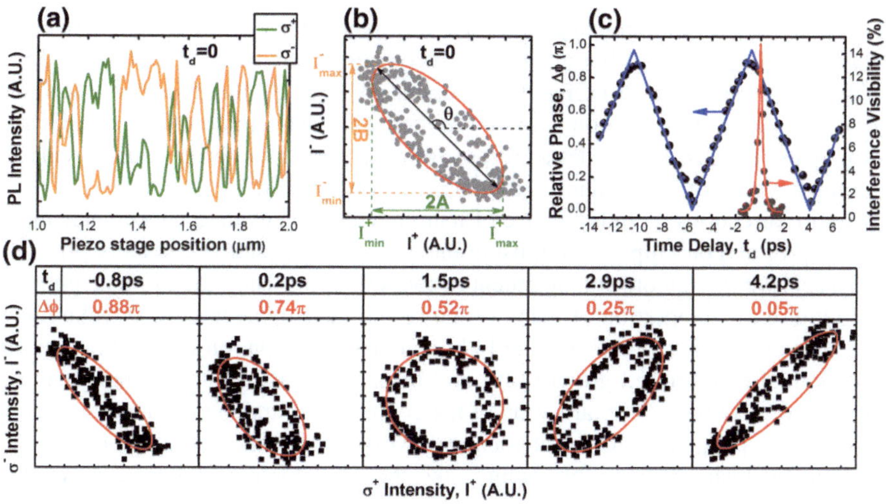

Fig. 4.12 **a** Interference fringes for σ^+ (*green*) and σ^- (*orange*) polarised light detected from O_T and O_R, recorded as the piezo mirror position is varied at $t_d = 0$. **b** Intensity plot of the interference data presented in **a** showing that the σ^+ and σ^- transitions are $\sim\pi$ out of phase at $t_d = 0$. The *red line* is the elliptical fit to the data which is used to extract the relative phase between the σ^+ and σ^- transitions. **c** The left hand axis shows a plot of $\Delta\phi$ as a function of the time delay. *The blue line* is a triangular waveform fit to the data. the right hand axis plots the visibility of the incoherent light interference of the QD and the fitting function (*red*) used to determine $t_d = 0$. **d** Intensity plots of the interference fringes recorded simultaneously from the σ^+ and σ^- transitions at different delay times of the interferometer. The *red lines* are the elliptical fits to the data

Figure 4.12c presents a plot of $\Delta\phi$ as a function of t_d. $\Delta\phi$ oscillates with a period of 10 ps, corresponding to the Zeeman splitting of 0.41 meV.

To determine $t_d = 0$ with high accuracy, the interference contrast of the entire QD ensemble, effectively a broadband light source, is measured and plotted in Fig. 4.12c. At zero time-delay, the relative phase between the σ^+ and σ^- lines is deduced to be 0.91π, close to the expected value of π, the difference in the relative phase between the x and y polarised components of $\sigma^\pm = (x \pm iy)$ light.

We note that for a path delay of $\lambda/4$, the σ^+ and σ^- lines will exit via opposite ports of the fiber beamsplitter, which can be implemented for in-plane read-out of the spin up/down state of the QD.

The above experiments demonstrate the polarisation of the photon emitted by the QD, including the relative phase between the polarisation components, and hence the full spin-state of the QD, is coherently mapped to a which-path state carried by the orthogonal waveguides.

4.8 Summary

We have presented a scheme for interfacing an optically addressed QD spin qubit to a path encoded photon using a crossed waveguide device. The polarisation state of a circularly polarised photon has been recovered at a remote location using an in plane interferometric measurement. This opens up the possibility of a highly scalable quantum optical network compatible with conventional semiconductor processing techniques. This work provides a promising way forward for constructing a scalable on-chip network of spins connected by optical waveguides.

Future directions include the use of nano-cavities to enhance light-matter interactions [28] and on-chip read-out using integrated single photon detectors [9].

To achieve the remote entanglement of two QD spins using the in-plane photons will require the integration of many of the recent developments in QD technology, such as nanocavities to enhance the light matter interaction, on-chip single photon detection, site-controlled growth [29] and resonant excitation that simultaneously allows access to the full polarisation of the control laser and the in-plane resonance fluorescence [30]. Despite the challenges involved, the rapid development of QD physics and technology to this point make the development of such a network a realistic and achievable goal.

References

1. A. Imamoğlu, D.D. Awschalom, G. Burkard, D.P. DiVincenzo, D. Loss, M. Sherwin, A. Small, Quantum information processing using quantum dot spins and cavity QED. Phys. Rev. Lett. **83**, 4204–4207 (1999)
2. J.L. OBrien, A. Furusawa, J. Vučković, Photonic quantum technologies. Nat. Photonics **3**, 687–695 (2009)

3. E. Knill, R. Laflamme, G.J. Milburn, A scheme for efficient quantum computation with linear optics. Nature **409**, 46–52 (2001)
4. A. Politi, J.C.F. Matthews, J.L. O'Brien, Shor's quantum factoring algorithm on a photonic chip. Science **325**, 1221 (2009)
5. A. Politi, M.J. Cryan, J.G. Rarity, S. Yu, J.L. O'Brien, Silica-on-silicon waveguide quantum circuits. Science **320**, 646–649 (2008)
6. Z. Yuan, B.E. Kardynal, R.M. Stevenson, A.J. Shields, C.J. Lobo, K. Cooper, N.S. Beattie, D.A. Ritchie, M. Pepper, Electrically driven single-photon source. Science **295**, 102–105, (2002)
7. C.L. Salter, R.M. Stevenson, I. Farrer, C.A. Nicoll, D.A. Ritchie, A.J. Shields, An entangled-light-emitting diode. Nature **465**, 594–597, (2010)
8. A. Dousse, J. Suffczyński, A. Beveratos, O. Krebs, A. Lemaître, I. Sagnes, J. Bloch, P. Voisin, P. Senellart, Ultrabright source of entangled photon pairs. Nature **466**, 217–220 (2010)
9. J.P. Sprengers, A. Gaggero, D. Sahin, S. Jahanmirinejad, G. Frucci, F. Mattioli, R. Leoni, J. Beetz, M. Lermer, M. Kamp, S. HoÌĹfling, R. Sanjines, A. Fiore, Waveguide superconducting single-photon detectors for integrated quantum photonic circuits. Appl. Phys. Lett. **99**(18), 181110 (2011)
10. R. Gingrich, P. Kok, H. Lee, F. Vatan, J. Dowling, All linear optical quantum memory based on quantum error correction. Phys. Rev. Lett. **91**, 1–4 (2003)
11. I. Fushman, D. Englund, A. Faraon, N. Stoltz, P. Petroff, J. Vuckovic, Controlled phase shifts with a single quantum dot. Science **320**, 769–72 (2008)
12. P. Borri, W. Langbein, S. Schneider, U. Woggon, R. Sellin, D. Ouyang, D. Bimberg, Ultralong dephasing time in InGaAs quantum dots. Phys. Rev. Lett. **87**, 1–4 (2001)
13. A.J. Bennett, R.B. Patel, J. Skiba-Szymanska, C.A. Nicoll, I. Farrer, Giant stark effect in the emission of single semiconductor quantum dots. Appl. Phys. Lett. **97**(3), 031104 (2010)
14. A. Laucht, F. Hofbauer, N. Hauke, J. Angele, S. Stobbe, M. Kaniber, G. Böhm, P. Lodahl, M.-C. Amann, J.J. Finley, Electrical control of spontaneous emission and strong coupling for a single quantum dot. New J. Phys. **11**(2), 023034 (2009)
15. J.P. Reithmaier, G. Sek, A. Löffler, C. Hofmann, S. Kuhn, S. Reitzenstein, L.V. Keldysh, V.D. Kulakovskii, T.L. Reinecke, A. Forchel, Strong coupling in a single quantum dot-semiconductor microcavity system. Nature **432**, 197–200 (2004)
16. T. Yoshie, A. Scherer, J. Hendrickson, G. Khitrova, H.M. Gibbs, G. Rupper, C. Ell, O.B. Shchekin, and D.G. Deppe, Vacuum Rabi splitting with a single quantum dot in a photonic crystal nanocavity. Nature **432** 200–203 (2004)
17. R.B. Patel, A.J. Bennett, I. Farrer, C.A. Nicoll, D.A. Ritchie, A.J. Shields, Two-photon interference of the emission from electrically tunable remote quantum dots. Nat. Photonics **4**, 632–635 (2010)
18. E.B. Flagg, A. Muller, S.V. Polyakov, A. Ling, A. Migdall, G.S. Solomon, Interference of single photons from two separate semiconductor quantum dots. Phys. Rev. Lett. **104**, 1–4 (2010)
19. A. Greilich, D.R. Yakovlev, A. Shabaev, A.L. Efros, I.A. Yugova, R. Oulton, V. Stavarache, D. Reuter, A. Wieck, M. Bayer, Mode locking of electron spin coherences in singly charged quantum dots. Science **313**, 341–345 (2006)
20. D. Brunner, B.D. Gerardot, P.A. Dalgarno, G. Wüst, K. Karrai, N.G. Stoltz, P.M. Petroff, R.J. Warburton, A coherent single-hole spin in a semiconductor. Science **325**, 70–72 (2009)
21. D. Press, T.D. Ladd, B. Zhang, Y. Yamamoto, Complete quantum control of a single quantum dot spin using ultrafast optical pulses. Nature **456**, 218–221 (2008)
22. T. Godden, J. Quilter, A. Ramsay, Y. Wu, P. Brereton, S. Boyle, I. Luxmoore, J. Puebla-Nunez, A. Fox, M. Skolnick, Coherent optical control of the spin of a single hole in an InAs/GaAs quantum dot. Phys. Rev. Lett. **108**, 1–5 (2012)
23. A. Berthelot, I. Favero, G. Cassabois, C. Voisin, C. Delalande, P. Roussignol, R. Ferreira, J.M. Gérard, Unconventional motional narrowing in the optical spectrum of a semiconductor quantum dot. Nat. Phys. **2**, 759–764 (2006)
24. A. Faraon, I. Fushman, D. Englund, N. Stoltz, J. Vučković, Dipole induced transparency in waveguide coupled photonic crystal cavities. Opt. Express **16**(16), 944–947 (2008)

25. M. Davanço, M.T. Rakher, D. Schuh, A. Badolato, K. Srinivasan, A circular dielectric grating for vertical extraction of single quantum dot emission. Appl. Phys. Lett. **99**(4), 041102 (2011)
26. C. Santori, D. Fattal, J. Vucković, G.S. Solomon, Y. Yamamoto, Indistinguishable photons from a single-photon device. Nature **419**, 594–597 (2002)
27. S. Laurent, S. Varoutsis, L. Le Gratiet, A. Lemaître, I. Sagnes, F. Raineri, A. Levenson, I. Robert-Philip, I. Abram, Indistinguishable single photons from a single-quantum dot in a two-dimensional photonic crystal cavity. Appl. Phys. Lett. **87**(16), 163107 (2005)
28. K. Rivoire, S. Buckley, J. Vučković, Multiply resonant photonic crystal nanocavities for non-linear frequency conversion. Opt. Express **19**, 22198 (2011)
29. A. Mohan, M. Felici, P. Gallo, B. Dwir, A. Rudra, J. Faist, E. Kapon, Polarization-entangled photons produced with high-symmetry site-controlled quantum dots. Nat. Photonics **4**, 302–306 (2010)
30. A. Muller, E.B. Flagg, P. Bianucci, X.Y. Wang, D.G. Deppe, W. Ma, J. Zhang, G.J. Salamo, M. Xiao, C.K. Shih, Resonance fluorescence from a coherently driven semiconductor quantum dot in a cavity. Phys. Rev. Lett. **99**, 2–5 (2007)

Chapter 5
Direct In-plane Readout of QD Spin

5.1 Introduction

The results presented in Chap. 4 describe a crossed waveguide device in which the two transitions of a Zeeman split doublet located at the intersection couple approximately equally to all four waveguides. This is the expected behaviour of the device in which in-plane polarisation transfer was demonstrated by reconstruction of the full spin state by combining the output from two orthogonal waveguides. This device will be henceforth referred to as Device-A.

A second device of the same design, which we will refer to as Device-B, fabricated on the same sample demonstrates markedly different behaviour. The σ^+/σ^- transitions of a Zeeman split doublet are directly mapped to anti-parallel waveguides. This provides direct in-plane spin read-out without the requirement of the off-chip interferometry that was used in Chap. 4. This is a surprising result for a non-chiral structure.

This chapter presents the data demonstrating direct in-plane QD spin readout, before discussing FDTD simulations of the structure. The simulations were conducted by our collaborators from the University of Bristol, R. Oulton and A. Thijssen. The simulations reproduce the behaviour of both devices by controlling the position of the QD within the intersection of the structure. A more intuitive explanation of the behaviour observed in the devices is then presented.

Having observed two distinct regimes of operation that are dependent on the location of the QD (direct read-out of spin information and in-plane transfer) we conclude that the way in which a spin optically couples to a photonic device is sensitive to the QD position and is therefore an important design consideration for any spin-photon interface.

The applications for a system with direct read-out of spin information include a deterministic-photon-path single-photon-source. Initial results using a spin-memory pumping technique investigating such a device are presented at the end of the chapter.

N. A. Wasley, *Nano-photonics in III–V Semiconductors for Integrated Quantum Optical Circuits*, Springer Theses, DOI: 10.1007/978-3-319-01514-9_5,

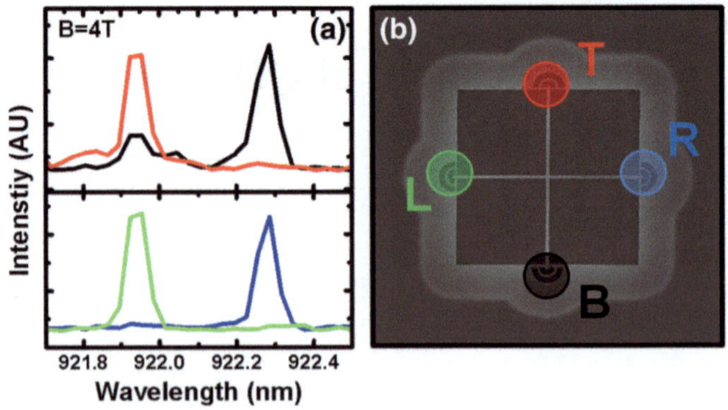

Fig. 5.1 **a** PL spectra of a Zeeman split QD doublet ($B = 3T$) recorded from the colour coded out-couplers on a crossed waveguide structure schematically represented in **b**

5.2 Experimental Observation of Direct In-plane Readout of QD Spin

Figure 5.1a, b show PL spectra recorded for a Zeeman split doublet recorded from the four out-couplers of Device-B using the same spatially selective optical set-up described in Sect. 2.7, with $B = 4T$. When collecting PL from the out-couplers, linear components of the $\sigma^+(\sigma^-)$ polarised transitions are observed from O_R and O_B (O_L and O_T).

The contrasts extracted from Fig. 5.1a where $C_\alpha = (I^+ - I^-)/(I^+ + I^-)$, $I^+(I^-)$ is the PL intensity of the $\sigma^+(\sigma^-)$ transition and $\alpha = L, R$ are $C_R = 0.92$ and $C_L = -0.93$, which corresponds to the direct read-out of the spin state of the QD in an in-plane geometry. This observation that the σ^+ transition is only observed from the top and right outcouplers whilst the σ^- transition is only observed from the bottom and left outcouplers is a surprising result for a symmetrical, non-chiral structure, with many possible applications in quantum optical circuits and networks.

To confirm this direct read-out requires a number of further experimental demonstrations.

The first is to demonstrate the two transitions originate from the same single QD. This proceeds with the observation that the two transitions are degenerate bar a fine structure splitting (FSS) at $B = 0T$. Figure 5.2a displays the PL spectra from the O_R and O_T out-couplers as the B-field is varied from $B = -4T$ to $B = 4T$. This shows that the two transitions are split by just 5 μeV at $B = 0T$, which is consistent with the fine structure splitting of an exciton complex as shown in Fig. 5.2b.

The PL spectra displayed in Fig. 5.2a also shows that the same behaviour occurs irrespective of the applied magnetic field. The same transitions are observed from the same out-coupler.

To confirm that it is a single QD which is responsible for producing both transitions a second-order cross-correlation measurement between light emitted from output

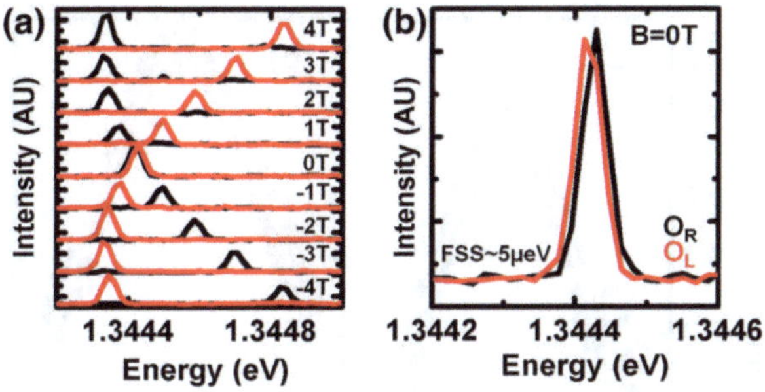

Fig. 5.2 **a** B-field dependent spectra recorded from outcouplers O_R (*black*) and O_L (*red*), in the range $B = -4T$ to $B = 4T$. At $B = 0T$ the spectra is displayed again in **b**. The two transitions have a fine structure splitting of ~5 μeV

couplers O_T and O_R is performed. The QD emission from the two transitions is filtered from other unwanted PL, using narrow band interference filters (FWHM of ~1.5 nm). The filters allow the spectrometer and fiber beam-splitter to be omitted from the optical set-up so that the transitions may be recorded directly by an APD, improving the incident rate. However, as the spectral filtering is much broader than what can be achieved using a spectrometer a high level of background signal is also recorded.

To correct for the large background signal we employ a correction to the raw data based on the estimated background signal, such that $g^2(\tau) = [C_N(\tau) - (1 - \rho^2)]/\rho^2$, where $C_N(\tau)$ is the normalized data and ρ is the signal to background ratio [1]. The background intensity is estimated by recording the intensity from the sample spatially detuned from the location of the QD. The cross-correlation histogram for the QD studied in Device-B is shown in Fig. 5.3a and shows anti-bunching with $g^{(2)}(0) = 0.04$, confirming the single photon nature of the emission.

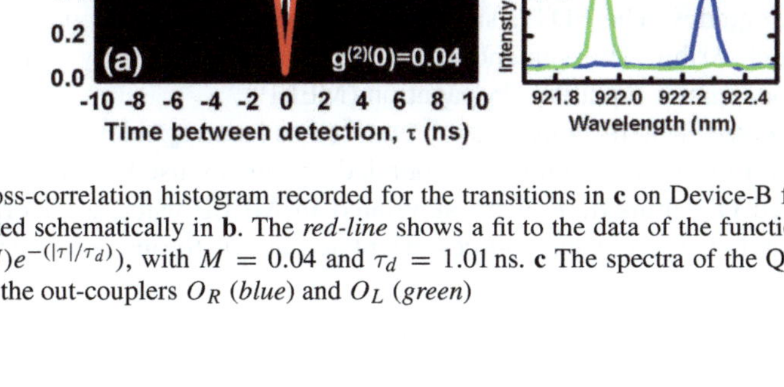

Fig. 5.3 Cross-correlation histogram recorded for the transitions in **c** on Device-B from O_R and O_L as depicted schematically in **b**. The *red-line* shows a fit to the data of the function $g^{(2)}(\tau) = (1 - (1 - M)e^{-(|\tau|/\tau_d)})$, with $M = 0.04$ and $\tau_d = 1.01$ ns. **c** The spectra of the QD transitions observed for the out-couplers O_R (*blue*) and O_L (*green*)

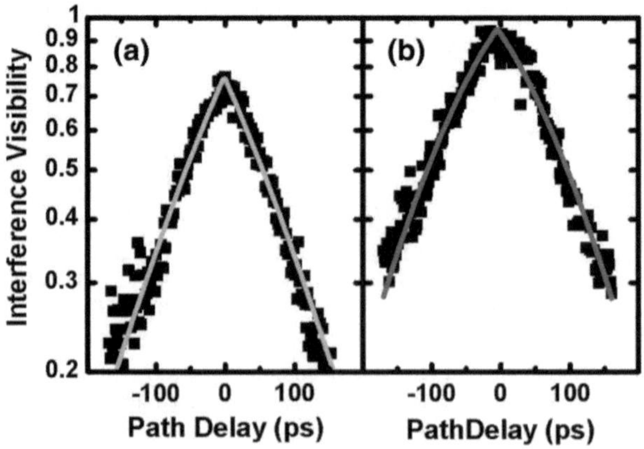

Fig. 5.4 Coherence time measurements of the **a** σ^+ and **b** σ^- transitions, respectively. The solid *red* and *blue* lines show fits to the data following Ref. [2]. Coherence times of 103 ps and 104 ps are found for the σ^+ and σ^- transitions, respectively

This is an experiment of note as it uses photons collected from two separate remote positions to conduct a cross-correlation measurement on a single QD. This represents progress towards the fabrication of a quantum optics circuit capable of performing a HBT type measurement on a single QD. The next stage would be to replace the out-couplers that terminate the waveguides with on-chip single photon detectors.

The coherence time for the QD in Device-B can be determined with Fourier transform spectroscopy by analysing the visibility decay curves fitted following Ref. [2]. From fits to the data, coherence times of 103 ps and 104 ps are found for the σ^+ and σ^- transitions, respectively, (Fig. 5.4). The measured coherence times of the two QDs are typical for InGaAs QDs in photonic structures [3, 4] and similar to the values observed in Device-A.

5.3 FDTD Simulation of Crossed-waveguide Structure

Finite difference time domain (FDTD) simulations were performed to investigate the differing properties observed in Devices A and B. The FDTD simulations were used to investigate how the exact location of the QD affects the coupling efficiency into the four waveguides. The FDTD simulations were performed by our collaborators on this project, Arthur Thijssen and Ruth Oulton at the University of Bristol, using MIT Electromagnetic Equation Propagation (MEEP).

The simulated system consisted of two crossed waveguides modelled as two 140×200 nm GaAs waveguides surrounded by air, as used in the experiments. The polarised QD transitions were simulated in MEEP using a superposition of two broadband charge-current sources orientated at 90° to one another with dipole

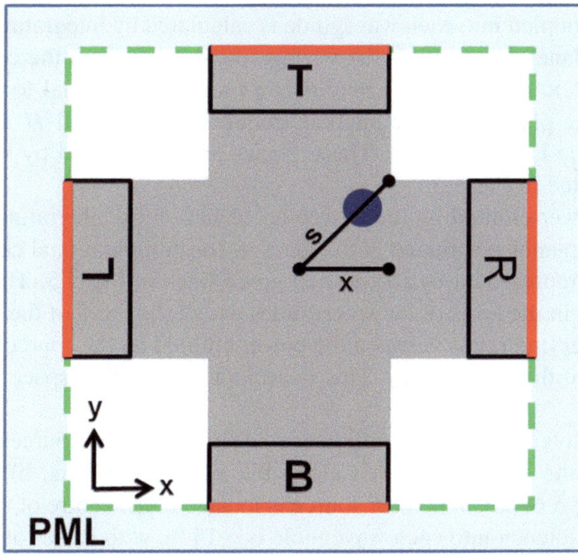

Fig. 5.5 Schematic diagram of the computational area of the waveguide intersection used in the FDTD simulations. The source is indicated by a *blue dot* and is located within the waveguide intersection at a position along a direction, s, diagonal from the centre or x, orthogonal to the y direction waveguide. The planes over which the Poynting vector is integrated to determine the power coupled into the waveguides are coloured *red*. The *green dashed line* represented the planes around the computational cell on the boundary with the perfectly matched layer (PML) over which the Poynting vector is integrated to determine the total power output of the source

moments in the orthogonal x and y directions. This is a standard technique within the MEEP framework to produce a source with an arbitrary polarisation that can be defined by controlling the amplitude, E and phase of the x (θ) and y (φ) dipoles according to $E_s = E_x \cos\theta/2 + E_y e^{i\varphi} \sin\theta/2$ for $0 \leq \theta \leq \pi$ and $0 \leq \varphi \leq 2\pi$.

The circularly polarised dipole source is located in the intersection at a distance s from the centre along the diagonal, or at a distance x from the centre perpendicular to the y direction waveguide, as represented in Fig. 5.5.

To prevent reflections, the waveguides run into a perfectly matched layer (PML).

5.3.1 Waveguide Coupling of QD Emission

The coupling efficiency of the source into each waveguide is calculated as a function of the source position as it is moved in the x and s directions away from the centre of the intersection. The position and polarisation dependent coupling of the QD to the cross waveguide structure is determined by calculating the power transmitted into the four waveguides.

The power coupled into each waveguide is calculated by integrating the Poynting vector over a plane at the exit of the waveguide, 4.2 μm from the centre: $P(\omega) = Re\{\hat{n} \cdot E(r,\omega)^* \times H(r,\omega)d^2r\}$ where $\hat{n}$ is a unit vector normal to the plane, r is a position in the plane, ω is the angular frequency and E and H are the Fourier transforms of the E and H-fields. These planes are represented by red lines in the diagram presented in Fig. 5.5.

The total power emitted by the source is calculated by integrating the Poynting vector over six planes positioned at the faces of the computational cell (the volume by simulation) represented by the dashed green lines in Fig. 5.5. The ratios of the fluxes collected in the arms of the waveguides and at the faces of the computational cell give the coupling ratios between the power emitted by the source and the power that couples into the waveguides. This is evaluated for a free-space wavelength of 922 nm.

Figure 5.6 plots the coupling efficiency of $\sigma^{\pm}$ polarised sources into the four waveguides as the source is moved along the x axis (Fig. 5.6a, b) and along the diagonal, s (Fig. 5.6c, d). When the source is located at the centre of the intersection the coupling efficiency into each waveguide is ~14 %, with losses of ~44 %.

The ideal coupling into each waveguide would be 25 %. Improvements to the efficiency of coupling can be made via optimisation of the waveguide dimensions which at present were chosen to ensure single mode propagation. However, further simulations of the device have suggested that the single mode operation of the nanowires would be maintained for waveguides wider than the current 200 nm. The simulations suggests an optimal design could reduce losses to below 20 %.

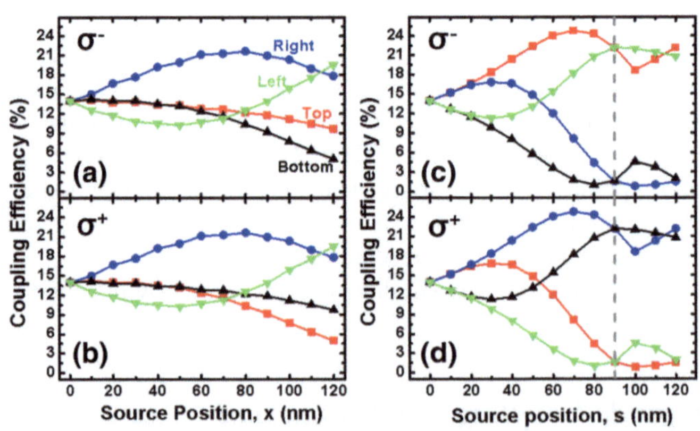

Fig. 5.6 FDTD calculations of the coupling efficiency for $\sigma^{\pm}$ polarised sources at different positions within the waveguide intersection. **a, b** σ^{+} and σ^{-} polarised source, respectively, shifted along the x axis. **c, d** σ^{+} and σ^{-} polarised source, respectively, shifted along the s direction at 45° to x. The *dotted grey line* at $s = 90$ nm indicates the inferred position of the QD in Device-B

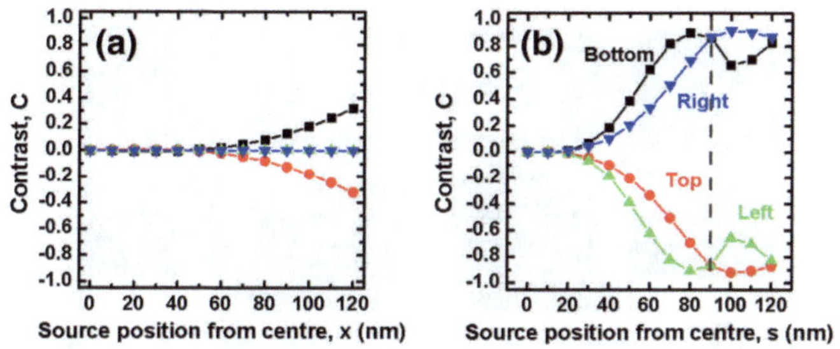

Fig. 5.7 Calculated contrast, ($C_\alpha = (I_+ - I_-)/(I_+ + I_-)$, where $\alpha = T, B, R, L$) between coupling of σ^+ and σ^- polarised light to the four waveguides as the position of the source is moved **a** horizontally along x and **b** diagonally along s. The *dotted line* at $s = 90$ nm indicates the inferred position of the QD in device-B

5.3.2 Position and Polarisation Dependent Contrast

To make a comparison with the experimental results the contrast of the linear component of the $\sigma^\pm$ polarised transitions in each waveguide where $C_\alpha = (I_+ - I_-)/(I_+ + I_-)$, and $\alpha = T, B, R, L$, is calculated. Figure 5.7 shows the contrast calculated for all four waveguides for different source positions.

It is clear from Fig. 5.7 that the experimental observations for Device-B are explained by a QD which is offset from the centre and in particular along the diagonal, s, direction rather than in x or y. In comparison the experimental observations of Device-A where the $\sigma^\pm$ polarised transitions couple equally to all waveguides is well described by a QD located at or close to (<20 nm) the centre of the intersection.

When the source is moved along x, there is no change in C for the right or left waveguides and only a small change in C for the top and bottom waveguides with a maximum $C_B = -C_R = 0.32$, when the source is moved 120 nm from the centre (Fig. 5.7a). However, when the source is moved along the diagonal, as shown in Fig. 5.7b, we observe a strong variation in C for all four waveguides and at $s = 90$ nm, the magnitude of C is approximately equal for all waveguides with $C_B = C_R = -C_T = -C_L = 0.87$, which is in good agreement with the experimental observations from Device-B.

This demonstrates that when the QD is located close to the centre, light is coupled equally into all four waveguides, as observed for Device-A. If however the QD is located at $s \approx 90$ nm, the linear components of the $\sigma+$ ($\sigma-$) transitions are directed along the bottom and right (top and left) waveguides, reproducing the experimentally observed behaviour of Device-B.

From the simulations, we infer that for Device-A, the QD is located $\lesssim$50 nm from the centre of the intersection, and the QD in Device-B, is located $s \approx 90$ nm. The operation of the two devices is illustrated by the electric-field distribution for light emitted by $\sigma^\pm$ polarised dipoles as presented in Fig. 5.8a–d for $s = 0$ and $s = 90$ nm respectively.

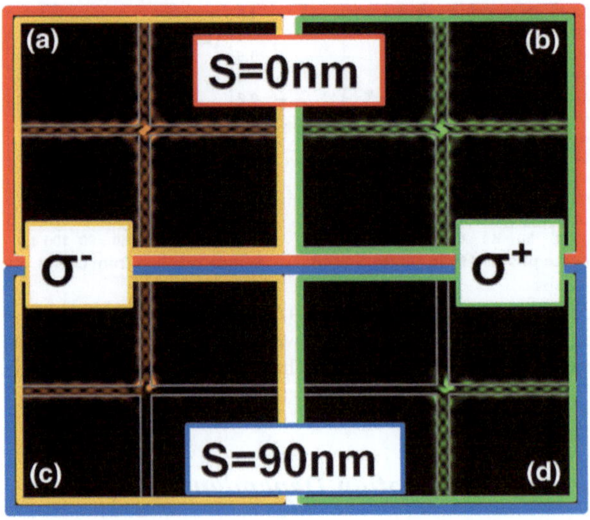

Fig. 5.8 Electric field intensity, $|E|^2$ at the centre of the waveguides for different source locations and polarisations 0.65 ps after the CW source begins to emit. **a** $\sigma-$ polarised source located at centre. **b** $\sigma+$ polarised source located at centre. **c** $\sigma-$ polarised source located off-centre at $s = 90$ nm. **d** $\sigma+$ polarised source located off-centre at $s = 90$ nm

5.3.3 *Effect of QD Position on the Fidelity of the Spin to Which-path Conversion*

An important role of the FDTD simulations is to verify that a one to one map between the polarisation of a photon emitted by a QD at the centre of the structure (Device-A) and the which-path state is achieved. That is to say, for any given source polarisation there exists a unique out-put path that isolates photons with a given polarisation. For example when photons with orthogonal polarisation are incident on a beam-splitter the photons will exit via orthogonal ports. By determining the polarisation basis in which this measurement is made, this isolation can be extended to any pair of orthogonal polarisations. The simulations also let us investigate whether Device-B performs a coherent map between polarisation and which-path state.

To determine this, the waveform of the $E_{x(y)}$ field in each waveguide, sampled at a point in the centre of the waveguide, equidistant from the centre of the intersection, is calculated for a QD at the centre, and an off-centre QD at $s = 90$ nm along the diagonal.

An amplitude, a and phase, φ are found by fitting the waveform to a cosine, and these are used to construct a phasor, $ae^{i\varphi}$. We then choose two of the waveguides (T, R, B, L), as defined in Fig. 5.9, to encode the which-path state, W for input polarisation, p as, $W(p) = a_1e^{i\varphi_1}, a_2e^{i\varphi_2}$.

If the device performs a one-to-one map between polarisation and the which-path state, then orthogonal input polarisations will map to orthogonal which-path

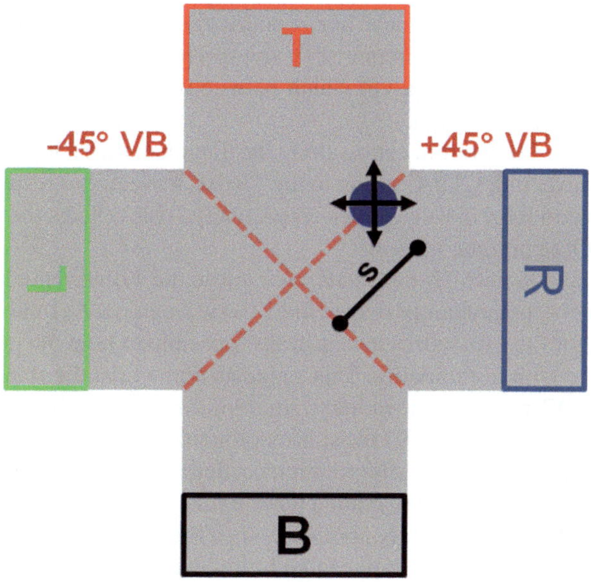

Fig. 5.9 The quantum dot is positioned a distance s from the centre along the diagonal. The intersection acts like a superposition of two 50:50 beam-splitters along the ±45° directions

states. To quantify this as a measure of the quality of the state conversion, the dot product, $O(WG : position) = W(p).W(p')^*$, of the which-path states arising from orthogonal input polarisations, p and p' is calculated. This is then normalised to half the total intensity out of all four waveguides. For an ideal one-to-one map $O(WG : position) = 0$.

To quantify the extent to which the spin-photon interface polarises the which-path state, we also calculate the total intensity of the light collected by the selected waveguide pair $I(WG : position)$, to check that this is independent of the input polarisation. The results are summarised in Table 5.1.

Table 5.1 Calculated measures of the quality of state conversion for the case of on and off-centre QDs

Polarisation basis	TR; centre		TB; off-centre	
	O (%)	I	O (%)	I
H/V	1.5	1.00/1.00	1.6	1.03/0.96
D/A	0.005	0.99/1.01	6.8	1.00/1.00
$\sigma+/\sigma-$	1.5	1.00/1.00	7	1.00/1.00

$O(WG; position)$ is the orthogonality, where a value of zero is ideal. $I(WG; position)$ is the total intensity, where a value of one is ideal. WG refers to the waveguide-pair used. The polarisation bases used are linear horizontal/vertical (H/V), linear diagonal/anti-diagonal (D/A) and $\sigma+/\sigma-$ circular

We use polarisation bases linear horizontal/vertical (H/V), linear diagonal/anti-diagonal (D/A), and right/left circular ($\sigma+/\sigma-$). The results presented in column 1 of Table 5.1, confirm that the spin to which-path conversion is near perfect for a QD at the centre of the intersection.

For a QD at $s = 90$ nm from centre along the diagonal, the H/V polarisations couple to all four waveguides, D/A couple to the TR/BL waveguides respectively, whilst $\sigma+/\sigma-$ couple to RB/LT waveguides respectively. This reflects the symmetries of the combined QD-photonic system.

To map the spin to a which-path state, we choose the TB waveguide-pair, since D and $\sigma+$ (A and $\sigma-$) polarisations map to the T (B) waveguide. As indicated in column 3 of Table 5.1, for the off-centre case, it is also possible to map the polarisation to a which-path state with high fidelity. This suggests that whilst the state conversion is sensitive to the QD position, the unitary nature of the map is robust.

Therefore for an off-centre QD the spin-photon map also works with high fidelity, albeit with an additional single qubit rotation needed to correct the spin to which-path map. Despite the different spin to guided photon maps resulting from QD location, it is possible to deduce any arbitrary spin state via a measurement of the intensity and phase at two output ports. This implies that following characterisation of the device, spin to guided-photon operation can be achieved regardless of QD position.

A key conclusion of this work is that the way in which a spin optically couples to a photonic device is sensitive to the QD position and is therefore an important design consideration for any spin-photon interface.

5.4 An Explanation of the Behaviour Based on Virtual Beam-splitters

The observations that a non-chiral photonic structure separates the $\sigma\pm$ components of the QD emission and breaks inversion symmetry between propagation in opposite directions are at first sight surprising. Although the FDTD simulations are capable of reproducing the experimental results by controlling the position of a QD within the intersection of the two waveguides it does not give an intuitive explanation of the behaviour that we see.

It is possible to provide a more intuitive description of the devices based on interpreting the device as a superposition of two virtual beam splitters orientated along the diagonal and anti diagonal axis. The following description of the waveguide intersection as a combination of virtual beam-splitters is due in part to analysis provided by A. J. Ramsay, who also made significant contributions to the work presented in Sect. 5.3.

An $\hat{x}$ polarised optical dipole will emit primarily into the k vectors $\pm k\hat{y}$ with a relative phase of π due to the fact that the source is a radiating dipole. This can be considered as inputs from the bottom and top waveguides respectively.

In principle the intersection can reflect or transmit the light into any of the four output ports. Therefore the function of the intersection can be simulated as a superposition of two identical virtual beam-splitters, with partially reflective surfaces along the $\pm 45°$ directions.

Figure 5.9 is a diagram of the waveguide intersection that shows the position of the QD in Device-B along the diagonal and the reflection planes of the two virtual beam-splitters.

The amplitudes of the four output ports for an $\hat{x}$-polarised dipole are calculated as a superposition of the four combinations of input ports and beam-splitter orientations and are displayed in Fig. 5.10.

In the case of the anti-diagonal beam-splitter, displacement of the source results in an additional position dependent phase shift. The amplitudes for the y-polarised dipole can be deduced along similar lines, and amplitudes for arbitrary polarised dipoles can be calculated as a superposition of the $\hat{x}$ and $\hat{y}$ dipoles.

The amplitudes can be derived as follows case (i) has amplitudes for T,R,B,L of $|\psi_{out}\rangle = (1; i; 0; 0)/\sqrt{2}$ as given in Ref. [5]. In case (ii), the input enters from the top port, which is an output in case (i). Hence the operation of the beam-splitter is the inverse operation of (i) and we take the complex-conjugate of the amplitudes. In addition the input has a π phase-shift, resulting in amplitudes $|\psi_{out}\rangle = (0; 0; -1; i)/\sqrt{2}$.

Case (iii) is equivalent to an input from port R in case (i), therefore the beam-splitter does the inverse operation and the reflection/transmission amplitudes are the complex conjugate of case (i). In addition the input has acquired a phase factor of e^{ika} due to the vertical displacement of the QD a, where $a = s/\sqrt{2}$, therefore the amplitudes are $|\psi_{out}\rangle = (e^{ika}; 0; 0; -ie^{ika})/\sqrt{2}$. Likewise case (iv) can be related to case (iii) using the same arguments going from case (i) to (ii) but with an additional overall phase factor of e^{-ika}.

This assumes that a beam injected into one of the waveguides is only weakly back-reflected and hence virtual beam-splitters along the $\hat{x}$-and $\hat{y}$-directions can be neglected. This has been confirmed experimentally and is true for reports of similar waveguide junctions [6].

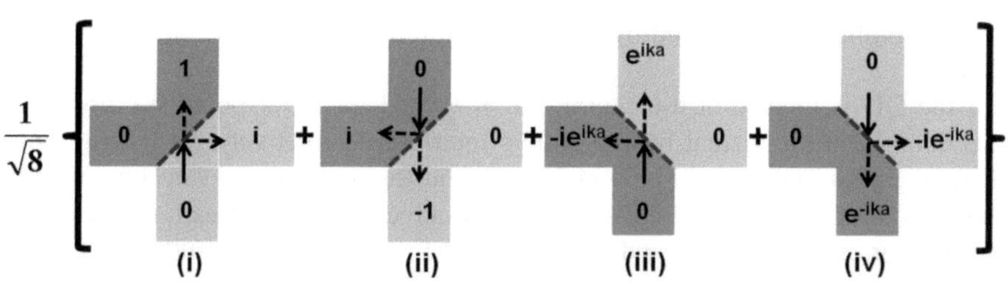

Fig. 5.10 The $\hat{x}$-polarised QD-dipole inputs light of wavevector $\pm k\hat{y}$. (*i, ii*) In the case of the $+45°$ beam-splitter, because the QD lies on the beam-splitter axis, these wave-vector components have identical input phase. (*iii, iv*) In the case of the $-45°$ beam-splitter, the QD lies above the beam-splitter axis and the different wave-vector components acquire a phase factor of $e^{\pm ika}$, where $a = s/\sqrt{2}$. The output is a superposition of the for possible input and beam-splitter configurations

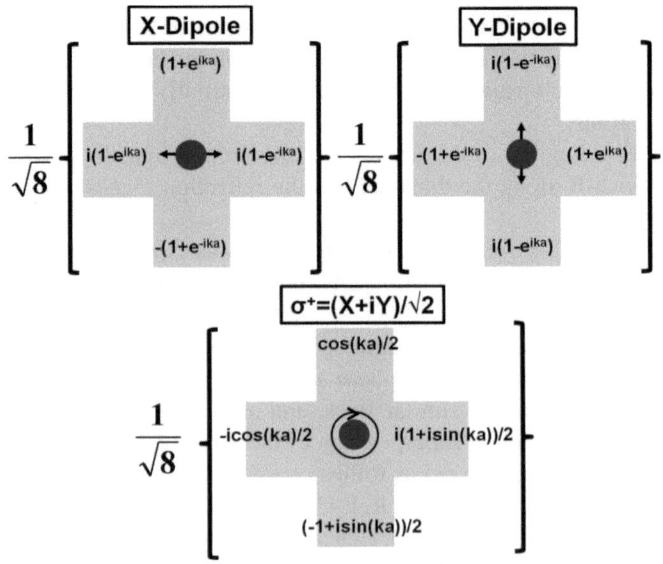

Fig. 5.11 (x-, y-Dipoles) These diagrams give the amplitudes of the output ports for the QD with $\hat{x}$- and $\hat{y}$- polarised optical dipoles respectively. The diagrams are mapped by a reflection about the $+45°$ axis. Note that in the case where the *dot* is at the centre, $ka = 0$, the x-polarised emission exits by the top and bottom waveguides only, as observed for Device-A. (σ^+) This gives the amplitudes of the output ports in the case of a σ^+ polarised optical dipole. This is found from an interference of an x and iy-polarised dipole. In the case where $ka = \pi/2$ the light exits from the right and bottom ports only, as observed for Device-B

Since there are no losses in this model the beam-splitters must be 50:50 to conserve energy. If the reflection and transmission coefficients are not equal, displacement of the source will change the total output power.

Figure 5.11 presents the resulting amplitudes of the output ports in the case of $\hat{x}$, $\hat{y}$, and σ^+ polarised optical dipole. The amplitudes of the $\hat{y}$-polarised dipole are found by reflecting the amplitudes of the $\hat{x}$-polarised dipole about the $+45°$-axis. The σ^+ amplitudes are found by adding the amplitudes of the x and iy-polarised dipoles.

If the QD is located at the centre, $ka = 0$, and for the $\hat{x}$-polarised dipole the reflections from the virtual beam-splitters interfere destructively for the right and left waveguides and the $\hat{x}$-dipole emits into the top and bottom waveguides only, as observed for Device-A. If however the QD is located off-centre along the diagonal, then for circularly polarised dipoles the T, L and B, R waveguides have equal QD-position dependent amplitudes, since a reflection about the diagonal maps the combined QD-photonic structure system from a σ^+ to σ^- dipole and vice-versa.

At $ka = \pi/2$ a destructive interference occurs and the σ^+ dipole is emitted into the right/bottom waveguides only as observed for Device-B. This occurs at $s = 92\,\text{nm}$ for $s = \frac{\lambda\sqrt{2}}{4n}$ using a refractive index of $n_{GaAs} \approx 3.5$, and $\lambda = 920\,\text{nm}$.

This interpretation supports the FDTD simulations which show that the contrast of $\sigma^{\pm}$ polarised light peaks at $s = 90$ nm, in close agreement to the value estimated using virtual beam-splitters.

In effect the displacement of the QD along the s direction adds a phase shift which leads to wave interference within the intersection that is responsible for the inversion symmetry of the crossed waveguide structure to be broken.

5.5 Summary

The scheme for interfacing an optically addressed spin qubit to a waveguide-photon using a crossed waveguide device presented in Chap. 4 has been further investigated in this chapter. We have observed the operation of this device in two regimes dependent on the location of the QD and shown that the device can be used for direct read-out of spin information as well as in-plane transfer.

In contrast to the more challenging goal of entangling two QD-spins by exploiting the in-plane spin transfer, the direct spin-readout has a number of interesting applications that are immediately accessible. These include the addition of integrated on-chip single photon detectors as terminators to the waveguides to replace the out-couplers currently used and spin-memory experiments where the QD is pumped into a single transition such that a subsequent photon path may be pre-determined.

5.6 Further Work: Spin-memory

By resonantly pumping a QD via the sharp resonance that occurs at a detuning of one longitudinal optical (LO) phonon above the QD state it has been shown that it is possible to pump a single transition of the Zeeman split doublet, based on the polarisation of the excitation laser [7]. This is due to coupling between the QD states and the surrounding crystal, this LO phonon mediated excitation offers a route to selectively prepare a QD spin state.

Spin pumping is an interesting effect in itself but also negates the need for transition resonant excitation in optical experiments to prepare a single QD spin state. This significantly simplifies a number of experimental requirements. One reason for this is that the LO-phonon resonance is broader than the QD transition, but also by simplifying the filtering of the excitation laser signal from the QD transition signal in collection that is required.

By employing this excitation technique we investigate Device-B and demonstrate a system in which the spin preparation technique pre-determines the emission path of the photon emitted from the QD.

Figure 5.12a displays the intensity of the Zeeman split ($B = 3T$) doublet transitions observed from the central intersection as the polarisation of the 850 nm excitation laser (which is approximately detuned from the 920 nm transitions by 2-LO

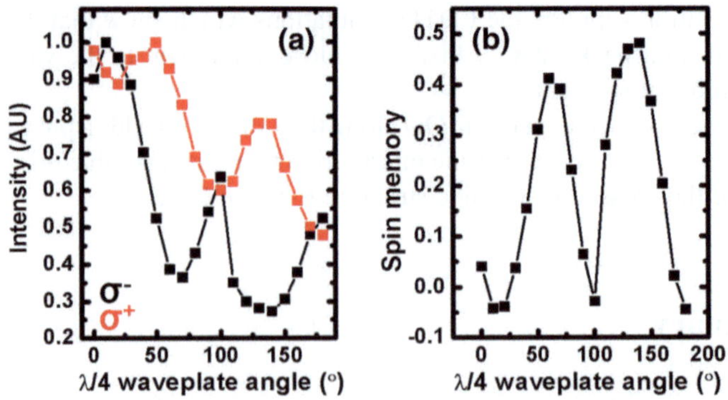

Fig. 5.12 **a** Intensity of the σ^+ (*red*), σ^-(*black*) Zeeman split ($B = 3T$) QD transitions studied in Device-B as the polarisation of the 850 nm excitation laser is modified by the rotation of a $\lambda/4$ waveplate in the excitation path. The PL from the QD transitions is collected from the same position as the excitation at the centre of the waveguide intersection. **b** The Spin memory $S = (I_{\sigma^+} - I_{\sigma^-})/(I_{\sigma^+} + I_{\sigma^-})$ of the σ^+, σ^- transitions as a function of the waveplate angle

phonons of energy ~35 meV) is modified by rotating a quarter waveplate in the excitation path.

The normalised intensity of the two transitions, I_{σ^+} and I_{σ^-} oscillate approximately out of phase. This is as expected for polarisation dependent preparation of the QD.

To quantify the fidelity of the preparation into a definite spin state we can define the spin memory $S = (I_{\sigma^+} - I_{\sigma^-})/(I_{\sigma^+} + I_{\sigma^-})$, the results are displayed in Fig. 5.12b. For ideal preparation the spin memory should oscillate between 1 and -1 describing a contrast between the intensity of emission from the σ^+ and σ^- transitions for the excitation laser polarisation of σ^+ and σ^-. In this case the spin memory oscillates between maximum values of 0.5 and -0.05.

This is good evidence that spin memory from the detuned excitation is occurring. There are a number of reasons why the fidelity is not unity. It is likely that the QD preparation into a single transition is not perfect and some excitation from another path other than the LO resonance is occurring as the excitation is approximately 2-LO phonon detuned. The location of the QD slightly off-centre of the waveguide cross section in a complicated photonic structure may also be compromising the polarisation of excitation experienced by QD and the collection efficiency of PL from the QD.

By performing the same experiment but collecting the QD PL from outcouplers, O_R and O_L, not only is the collection efficiency problems associated with the complicated structure at the centre omitted but the direct path control over photon emission can also be demonstrated.

Figure 5.13a displays the spin memory for central excitation and out-coupler collection $S = (I_L - I_R)/(I_L + I_R)$. The spin memory oscillates with the quarter waveplate angle, as was seen with the collection from the waveguide intersection.

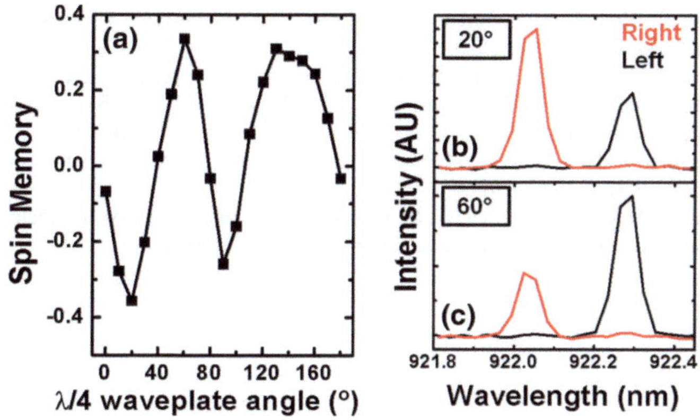

Fig. 5.13 **a** Spin memory as recorded from collection at out-couplers O_R and O_L, where $S = (I_L - I_R)/(I_L + I_R)$. **b** Recorded spectra when $S_{min} \sim -0.4$ at a waveplate angle of 60° and $S_{max} \sim 0.4$ at 20°

However, the spin memory is now more symmetrical about $S = 0$ with maximum values of $S_{min} \sim -0.4$ and $S_{max} \sim 0.4$. This result is a large improvement toward the ideal values.

The spectra recorded from the out-couplers that display the maximum control over the path control from the polarised excitation are shown in Fig. 5.13b. The spectra demonstrate that there is good control of the emitted photon path from this spin memory pumping technique.

The contrast of $C \sim 0.9$ observed for the σ^+, σ^- transitions from O_R to O_B suggest that it is the preparation of the spin state that is limiting the control over the deterministic excitation to which path relationship. Improving the spin-memory, S, would be enhanced by excitation at the 1-LO resonance but may require a resonant pumping technique [8], which will improve the QD preparation fidelity. However this technique will require both a tunable laser and improved spatial filtering in the collection path.

References

1. R. Brouri, A. Beveratos, J. Poizat, P. Grangier, Photon antibunching in the fluorescence of individual color centers in diamond. Opt. Lett. **25**(17), 1294–1296 (2000)
2. A. Berthelot, I. Favero, G. Cassabois, C. Voisin, C. Delalande, P. Roussignol, R. Ferreira, J.M. Gérard, Unconventional motional narrowing in the optical spectrum of a semiconductor quantum dot. Nat. Phys. **2**, 759–764 (2006)
3. C. Santori, D. Fattal, J. Vucković, G.S. Solomon, Y. Yamamoto, Indistinguishable photons from a single-photon device. Nature **419**, 594–597 (2002)
4. S. Laurent, S. Varoutsis, L. Le Gratiet, A. Lemaître, I. Sagnes, F. Raineri, A. Levenson, I. Robert-Philip, I. Abram, Indistinguishable single photons from a single-quantum dot in a two-dimensional photonic crystal cavity. Appl. Phys. Lett. **87**(16), 163107 (2005)

5. B.W. Lovett, P. Kok, *Introduction to Optical Quantum Information Processing* (Cambridge Uniiversity Press, Cambridge, 2010)
6. S.G. Johnson, C. Manolatou, S. Fan, P.R. Villeneuve, J.D. Joannopoulos, H.A. Haus, Elimination of cross talk in waveguide intersections. Opt. Lett. **23**, 1855–1857 (1998)
7. A. Boyer de la Giroday, A. Bennett, M. Pooley, R. Stevenson, N. Sköld, R. Patel, I. Farrer, D. Ritchie, A. Shields, All-electrical coherent control of the exciton states in a single quantum dot. Phys. Rev. B **82**, 2–5 (2010)
8. M. Kroutvar, Y. Ducommun, D. Heiss, M. Bichler, Optically programmable electron spin memory using semiconductor quantum dots. Nature **432**, 81–84 (2004)

Chapter 6
InP QDs in GaInP Photonic Crystal Cavities

6.1 Introduction

For certain applications the emission wavelength of InAs QDs (900–1300 nm) is not ideal, as it is far from the maximum efficiency of commercially available Si based detectors. InP QDs embedded in GaInP provide an alternative with the significant advantage of emitting in the red spectral range, at the maximum efficiency of Si detectors. This makes InP QDs an attractive proposition for implementation in optical quantum information processing and free space quantum communication [1].

A further advantage of InP QDs is that single photon emission has been demonstrated up to 80 K, opening up the possibility of a liquid nitrogen cooled single photon source [2]. PCs operating in the visible are also of interest for coupling to external light emitters, such as nitrogen vacancy centers in diamond nanocrystals [3] and fluorescent molecules [4].

Most of the research on PC cavities in this wavelength range has been conducted using SiN_4 membranes, where the Q-factor is limited by the low refractive index $n = 2$, with reported Q-factors of 1,500 for L3 cavities [4] and 3,400 for heterostructure cavities [5]. GaP [6] and GaInAlP [7] have also been used to fabricate L3 cavities with modes in the red, with maximum Q-factors of 1,700 and 2,000, respectively.

We have recently demonstrated that L3 PC cavities can be fabricated in GaInP, with Q-factors as high as 7,500 and, for the the first time, semiconductor QDs coupled to PC cavity modes operating in the visible spectrum [8]. Using time resolved photoluminescence measurements spontaneous emission suppression of the QDs within the photonic band gap was observed and for a resonantly coupled QD a Purcell enhancement of 8 was recorded.

This chapter will present and discuss aspects of this work before going on to discuss Fourier spectroscopy of InP QDs in GaInP photonic crystal cavities and how this technique is used to probe the decoherence processes that occur in InP QDs.

N. A. Wasley, *Nano-photonics in III–V Semiconductors for Integrated Quantum Optical Circuits*, Springer Theses, DOI: 10.1007/978-3-319-01514-9_6,

6.2 Sample

The wafer used in these studies was grown by metal organic vapour phase epitaxy on 3° misorientated (100) GaAs. It consisted of a 100 nm GaInP waveguide incorporating a p-i-n junction with a layer of InP QDs at its centre, grown on top of a 670 nm n-doped $Al_{0.6}Ga_{0.4}As$ sacrificial layer and an undoped substrate. This sample was intended for devices incorporating electric field tuning although this will not be discussed here. For the QD growth, 3.3 Å of InP was deposited at a growth rate of 1.1 Ås at 650 °C. The PCCs were fabricated using electron beam lithography and a $SiCl_4$ reactive ion etch. The suspended membrane was released by selectively wet etching the $Al_{0.6}Ga_{0.4}As$ layer using hydrofluoric acid. A SEM image of a photonic crystal used in these experiments is shown in Fig. 6.1a.

6.3 Experimental Set-up

The experiments were conducted using micro-photoluminescence (μPL) techniques in a helium flow cryostat at ∼10 K. Emission from the QDs was excited by focussing a 650 nm laser diode to a spot with a diameter of 1 μm with a 100 × ($NA = 0.75$) objective lens. The emitted light was collected through the same objective.

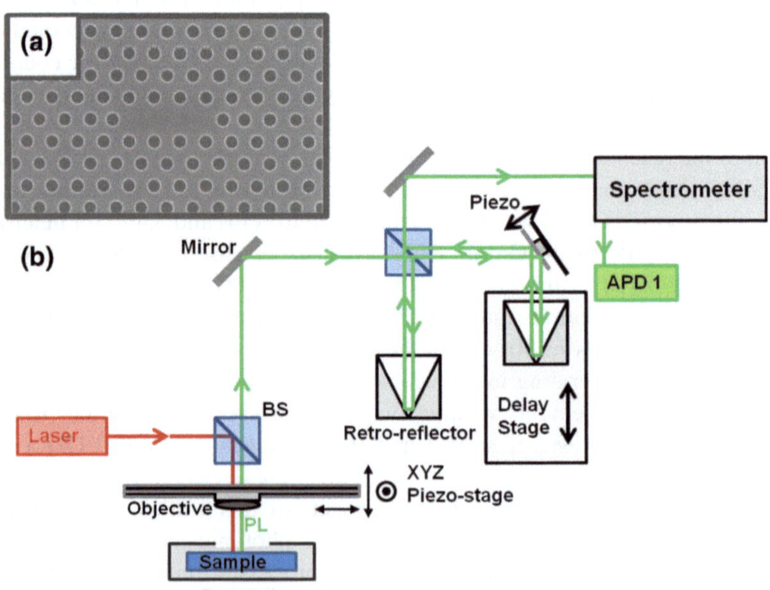

Fig. 6.1 **a** SEM image of an L3 type photonic crystal cavity used in these experiments. **b** Experimental set-up of the μPL and Michelson interferometer system

For the interference experiments, the collected light was directed into a free space Michelson interferometer consisting of a beam splitter and two retro-reflecting arms, as shown in Fig. 6.1b. The path length difference between the arms was controlled by a motorised delay stage and a piezo-actuated mirror for coarse and fine control respectively. The light was recombined and interfered on the initial beam splitter before being filtered with a 0.55 m single spectrometer and the intensity recorded on a Si based avalanche photo-diode (APD).

Hanbury–Brown and Twiss autocorrelation measurements were conducted with the addition of a fiber beam splitter, two Si APDs, and a single photon counting module with one arm of the Michelson interferometer blocked.

For time resolved lifetime measurements a 405 nm pulsed laser diode with a pulse width of ~50 ps was used to excite the QD emission and the signal recorded on a Si APD with a temporal resolution of ~40 ps resulting in a system response of ~100 ps.

6.4 Control of Spontaneous Emission of InP QDs in GaInP Photonic Crystal Cavities

Single quantum dots (QDs) integrated within photonic crystal cavities (PCC) are much studied systems because they provide an environment in which emission from an approximately two level quantum emitter (QD) may be confined for long periods of time (high Q) in a nanoscale volume. These properties can be used to modify many of the emission characteristics of the embedded QDs and provide a means of improving the extraction efficiency of QD emission [9–11]. As such, QDs embedded in PCCs are ideal platforms for studying cavity quantum electrodynamics whilst offering scalable functionality for many proposed quantum information processing schemes [12, 13].

Although the motivation to fabricate QD-PCC systems are common to all devices, a number of different cavity designs exist that provide flexibility in functionality (H1, L3, Heterostructure etc.). The properties of a QD-PCC system are also affected by the material system from which they are created. Typically InAs/GaAs material systems have been used in the development of QD-PCC systems in the last ten years, making many significant advances including the observation of strong-coupling [14]. However there is motivation to develop QD-PCCs from other material systems. In particular, as previously discussed, an InP/GaInP system would provide a number of benefits not currently available to a InAs/GaAs material system.

In this work L3 type photonic crystal cavities were fabricated in GaInP. By performing finite difference time domain simulations of the modified L3 defect cavity [15] the parameters required for fundamental mode emission in the region 670–700 nm were determined and optimized for maximum Q-factor. An SEM image of the fabricated PCC is displayed in Fig. 6.1a. With a lattice period $a = 187$ nm, hole radius $r = 0.3a$, and end hole shift $s = 0.18a$ Q-factors as high as 7,500 were achieved (the simulated Q-factor for this structure was 85,000). The recorded μPL

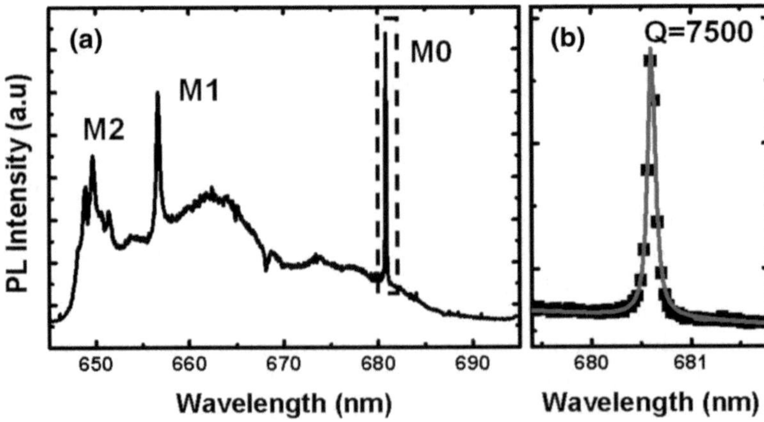

Fig. 6.2 **a** μPL spectra recorded from a typical GaInP L3 PC, showing emission from InP QD ensemble (650–685 nm) and cavity modes. **b** Fundamental mode emission of cavity with quality factor of ~7,500

spectra showing emission from the InP QD ensemble with the L3 cavity modes is displayed in Fig. 6.2a. The fundamental mode emission of the cavity with quality factor of ~7,500 is also displayed in Fig. 6.2b.

From μPL spectroscopy the decay dynamics of the QD emission were measured by selecting well isolated QD lines and conducting time resolved photoluminescence measurements. A typical lifetime of $\tau = 0.66$ ns was observed for QDs located in the unpatterned bulk of the sample and $\tau = 13.1$ ns when the QD was located in the patterned crystal and spectrally within the photonic band gap. This increase in lifetime results from the reduction in the density of states within the photonic band gap available for the QD transition.

A main motivation for fabricating PCCs is to control the spontaneous emission rate of QDs. For a QD located with the PCC and resonant to a PCC mode there is coupling between the cavity mode and the QD state. Depending on the strength of this interaction two regimes of coupling are defined: strong coupling and weak coupling. In the weak coupling regime a Purcell enhancement in the emission rate of the QD is expected due to an increase in the available DOS, as discussed in Chap. 1, Sect. 1.4.1.

Figure 6.3a shows a μPL spectrum in which several QD lines are observed spectrally close to the fundamental mode in a L3 type GaInP PCC. In this case, the mode has $Q = 3,300$. Here, we concentrate on the QD detuned from the mode by $\Delta\lambda = 0.56$ nm, where $\Delta\lambda = \lambda_{mode} - \lambda_{QD}$, labeled QD in Fig. 6.3a. By increasing the temperature $\Delta\lambda$ decreases until the QD comes into resonance with the mode at 44.5 K. Figure 6.3a also shows the μPL spectrum for the $\Delta\lambda = 0$ case.

Figure 6.3b shows that this tuning effects the QD lifetime, which decreases from ~0.75 ns when $\Delta\lambda = 0.56$ nm to ~90 ps at $\Delta\lambda = 0$. The decay transient is fitted with a biexponential decay,with the ~90 ps decay attributed to the Purcell enhanced

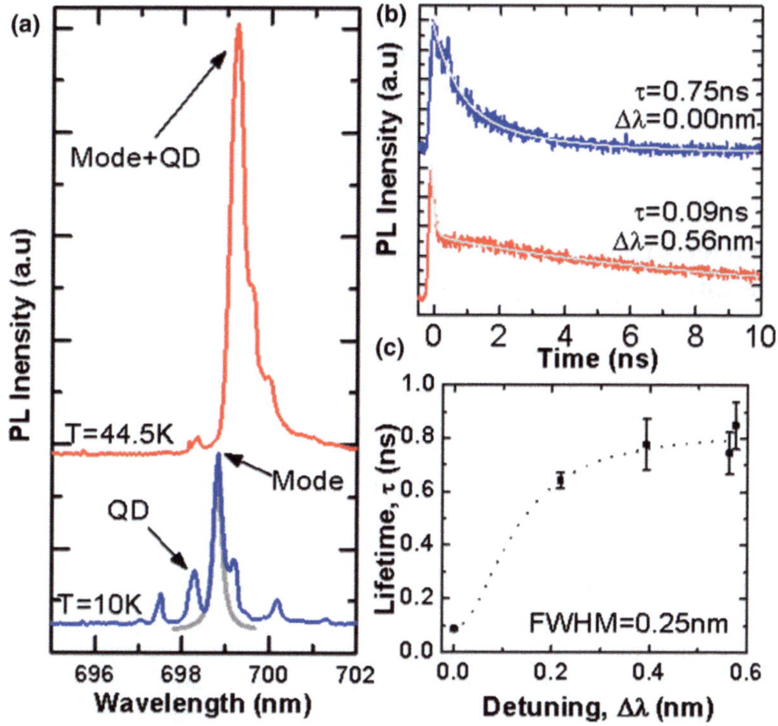

Fig. 6.3 **a** Photoluminescence spectra of a weakly coupled QD-cavity system recorded at 10 and 44.5 K. Light gray line shows a Lorentzian fit to the cavity mode resonance at 10 K, with Q = 3,300 and FWHM = 0.21 nm. **b** Decay transients recorded from the QD at 10 K ($\Delta\lambda = 0.56$ nm) and 44.5 K ($\Delta\lambda = 0$ nm). **c** Measured QD lifetime as a function of QD-cavity mode detuning. Dashed line shows Lorentzian fit, with FWHM = 0.25 nm

emission from the QD and a longer lived decay of ~9 ns attributed to nonresonant feeding of the mode from other QDs within the PC [16, 17]. The decrease in lifetime corresponds to a Purcell enhancement of ~8.

To verify that this increase in the spontaneous emission rate is due to the Purcell effect, the QD lifetime at various detunings, as shown in Fig. 6.3c was measured. The enhancement of the spontaneous emission rate follows the Lorentzian line shape of the mode as the QD is detuned from resonance [18]. The dashed line in Fig. 6.3c shows a Lorentzian fit to the lifetime measurements, with a FWHM = 0.25 nm, which is in good agreement with the measured linewidth of the mode and confirms that the reduction in lifetime results from the Purcell effect.

This observation of weak coupling between InP QDs and GaInP in L3 photonic crystal cavities, demonstrates that InP/GaInP QD-PCC systems are a possible route to a number of quantum optics and nano-photonic goals, with the benefits of emission in the 600–700 nm spectral range. One such goal is an on-chip indistinguishable single photon source. However, despite the advantageous emission properties, InP

QDs suffer from short coherence times that restrict their viability as a source of indistinguishable single photons, as discussed further in the next part of this chapter. An investigation into the decoherence processes of InP QDs in photonic crystal cavities in the context of achieving an InP/GaInP indistinguishable photon source is also presented.

6.5 Fourier Transform Spectroscopy of InP QDs in GaInP Photonic Crystal Cavities

Many proposed quantum optics experiments and quantum information processing schemes require reliable indistinguishable single photons [19]. By means of Fourier spectroscopy we observe the coherence times of InP QD emission from L3 type photonic crystal cavities and analyse the decoherence processes currently limiting the production of a PCC based source of single indistinguishable photons from InP QDs.

6.5.1 *QD Coherence*

The interaction between a QD and its environment destroys the coherence of photon emission with a typical dephasing time T_2^*. The fluctuations of the interaction cause spectral shifts in the emission energy over a range Σ between single quantum transitions rendering the photons distinguishable.

The coherence time of photons from the QD source, T_2, is defined as,

$$\frac{1}{T_2} = \frac{1}{2T_1} + \frac{1}{T_2^*} \tag{6.1}$$

where T_1 is the radiative lifetime of the source [20].

6.5.2 *Indistinguishability*

The indistinguishability ratio $R = \frac{T_2}{2T_1}$, quantifies the extent to which photons produced from a source are indistinguishable. In a case where the coherence time of a source is lifetime limited because no dephasing takes place, $T_2 = 2T_1$, the indistinguishability ratio reaches a maximum value of $R = 1$. Each photon produced from this source is indistinguishable from every other photon produced from the same source. The number of indistinguishable photons produced is only limited by the emission rate, T_1.

To increase R for a given source requires a reduction in T_1 or an increase in T_2^*. A reduction in T_1 may be achieved using an optical cavity to exploit the Purcell effect and enhance the spontaneous emission rate. However, if T_2 is dominated by

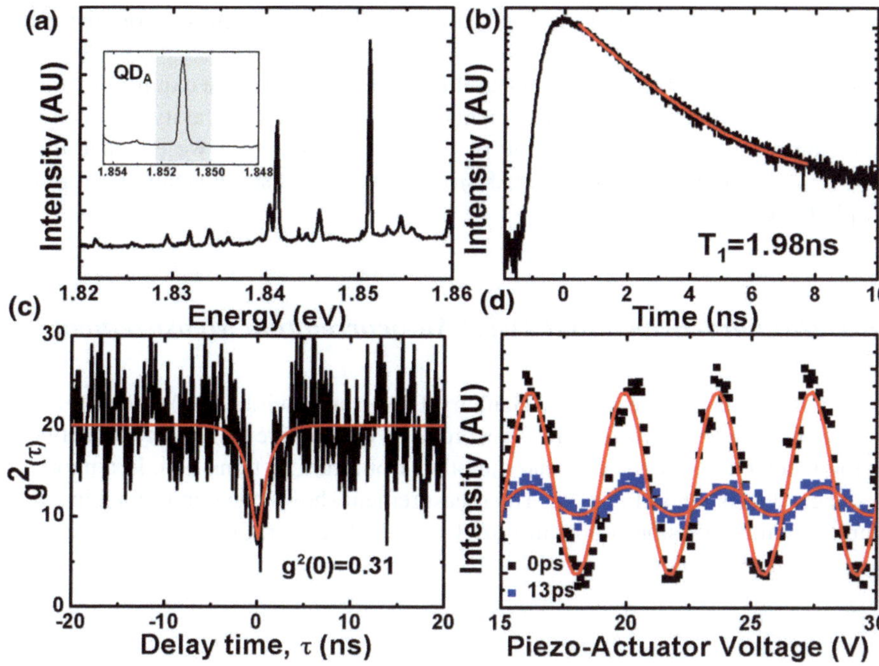

Fig. 6.4 **a** μPL spectra recorded from a typical L3 PC cavity showing InP QD emission. The inset shows a QD line at 1.851 eV (670 nm) used in the following experiments, QD_A, with the *grey shaded* area showing the spectral window used for interferometry. **b** Decay transient recorded for QD_A, *red line* is a monexponential fit to the data with a decay time $T_1 = 1.98$ ns. **c** HBT autocorrelation measurement from QD_A with a fitted $g^2(0) = 0.31$. **d** Interference fringe contrast measured as a function of piezo-actuator voltage for path delay 0 ps (*black*) and 13 ps (*blue*) and their respective fits (*red*)

T_2^* ($T_2^* \ll T_1$), T_2 is difficult to increase as it requires a significant reduction in the dephasing caused by intrinsic properties of the material, such as charge trapping sites and localised defects. As a result, previous attempts to improve the R from QDs have concentrated on using the Purcell enhancement provided by photonic modes to reduce T_1. These measurements have been made in InAs systems with best results achieving $T_1 \sim T_2^*$ [20, 21].

6.6 Characterisation

Figure 6.4a shows the μPL spectrum recorded at low power from QD emission in a PCC. Narrow lines corresponding to individual QD transitions are observed. Figure 6.4b shows an emission decay transient of an off resonant InP QD in a PCC. By fitting the decay transient with an exponential decay we observe a radiative decay lifetime $T_1 = 1.98$ ns, which is approximately similar to values reported for off resonant InP QDs [8].

To achieve a desirable indistinguishability value of $R \geq 0.5$, we require $T_2^* \geq 2T_1$. The largest reported Purcell enhancement of an InP QD in a PC cavity has been ~8, producing a radiative decay lifetime of 0.09 ns [8]. Therefore we estimate that a $T_2^* \geq 0.18$ ns is required to achieve a desirable level of indistinguishability from InP QD photon emission.

6.6.1 Hanbury-Brown and Twiss Autocorrelation Measurement

Figure 6.4c displays the results of a Hanbury-Brown and Twiss (HBT) autocorrelation measurement, where a histogram is recorded of the relative time delay, τ, between a photon detection at one APD and the other from the QD transition, in a photonic crystal cavity, shown in Fig. 6.4a. The measurement shows clear anti-bunching with a multi-photon emission probability $g^2(0) = 0.31$, confirming the InP QDs are a source of single photons.

6.7 Fourier Transform Spectroscopy

Fourier transform spectroscopy was performed to determine the coherence time, T_2, of individual QD lines. The signal intensity, I, for a spectral line as a function of the path delay time, t, in a Michelson interferometer, $I(t)$, is given by

$$I(t) = I_0 \left[1 + C(t) \cos \left(\frac{E_0 t}{\hbar} \right) \right] \tag{6.2}$$

where I_0 is the average intensity, E_0 is the central energy of the transition and $C(t)$ the interference visibility.

By modulating the path delay over a small number of wavelengths with the piezo-actuator, we can assume that $C(t)$ is constant over the small change in path delay time ($\Delta t \approx 11\ fs$). The function $I(t)$ may then be fitted to the data to extract an approximate value of C for a given t. A coarse path delay is then introduced using the motorised delay stage and the measurement repeated, allowing a trace of C to be plotted as a function of t. Figure 6.4d displays the intensity recorded on the APD for two traces taken at a coarse path delay of 0 ps and 13 ps, fitted with Eq. 6.2. The interference visibility, $C = (I_{max} - I_{min})/(I_{max} + I_{min})$, decays as the path delay increases.

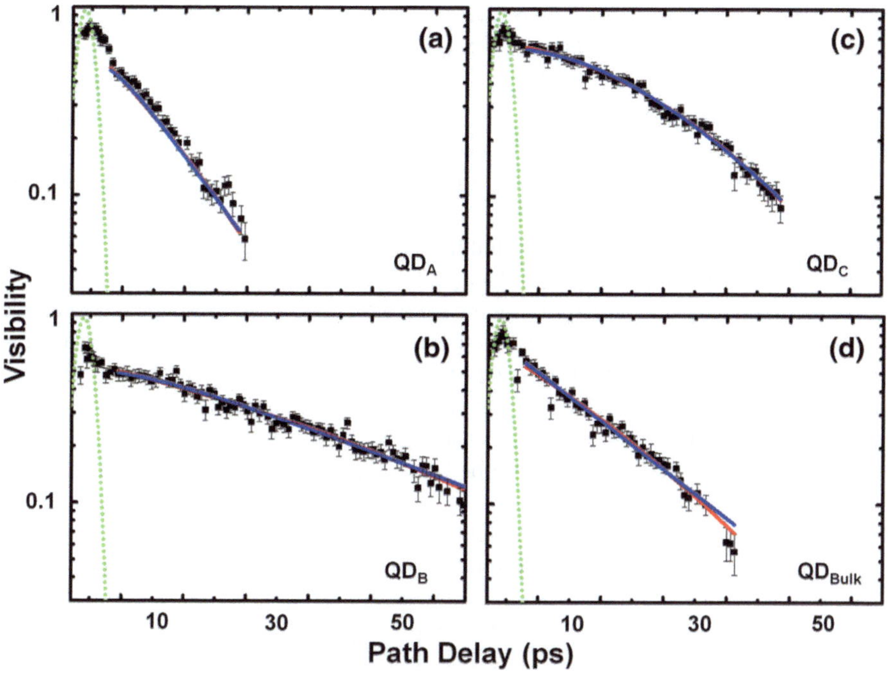

Fig. 6.5 Semi-log interference visibility traces for three QD transitions **a** QD_A **b** QD_B and **c** QD_C located in a photonic crystal cavity and a QD transition taken from an unpatterned area of the sample, **d** QD_{Bulk}. The data sets have been fitted with function $C_1(t)$ (*Blue*) and $C_2(t)$ (*Red*). The *dashed green line* represents the system response function for incoherent light obtained from *white* light illumination

6.7.1 Analysis

The process through which the coherence of a QD is lost is related to the fluctuations in the interaction between the QD and the environment in which the QD is located. The nature and timescale of these fluctuations determine both the coherence time and the lineshape of the emission line, which in turn affects the form of the visibility as a function of path delay time, $C(t)$.

Figure 6.5b displays the measured interference visibility as a function of path delay $C(t)$ on a semi-log plot for the emission line shown in Fig. 6.4a. The dashed green line represents the system response to incoherent light and defines the resolution limit of the system.

By applying the stochastic theory of lineshape and relaxation of a system coupled to a fluctuating reservoir the emission spectrum of a quantum dot is characterised in the time domain and, assuming Gaussian fluctuation, the Fourier-transform of the intensity spectrum has the general expression

$$C_1(t) = exp\left[\frac{-\Sigma^2\tau^2}{\hbar^2}\left(\exp\left(\frac{-t}{\tau}\right) + \frac{t}{\tau - 1}\right)\right] \tag{6.3}$$

where Σ is the spectral range over which the energy of emission is modulated and τ is the characteristic timescale of modulation [22]. When $\Sigma\tau \gg \hbar$ the function $C_1(t)$ tends toward a Gaussian decay profile and in the limit $\Sigma\tau \ll \hbar$ the function approaches a Lorentzian profile.

Two functions, $C_1(t)$ and $C_2(t)$, that have been reported elsewhere are used here to analyse the structure of the $C(t)$ data [23].

By using this approach of applying the stochastic theory of lineshape it is possible to distinguish between two causes of spectral lineshape broadening based on the variation in the characteristic timescale of fluctuations associated with each process, namely energy modulation due to trapped carriers [24] and dephasing processes due to exciton-phonon interactions [23]. (An acoustic phonon with energy ~1 meV in InP has a frequency of 240 GHz while trapped carrier processes have typical frequencies of around 1,000 Hz).

$C_1(t)$ can be expressed as a product of a Gaussian and Lorentzian decay due to trapped carriers and phonon interactions respectively

$$C_2(t) = exp\left(\frac{-\Gamma t}{\hbar}\right) exp\left(\frac{-\Delta^2 t^2}{2\hbar^2}\right) \tag{6.4}$$

where Γ is the product of the modulation energy due to a phonon and the timescale of phonon modulation and Δ is the modulation energy due to trapped carriers, the details of which arc described in Ref. [23].

In Fig. 6.5a–d visibility data recorded from four separate QDs have been fitted with the visibility functions $C_1(t)$ (blue) and $C_2(t)$ (red). Figure 6.5d is the visibility data recorded from a typical QD found in the bulk material outside of the PCC and is shown here for comparison to the QDs located within the photonic crystals. The shape and length of the visibility decays traces differ considerably between the four QDs.

The full width at half maximum (FWHM) ΔV for the overall linewidth of the QD line is estimated from the function $C_1(t)$ by determining whether the Gaussian ($\Sigma\tau \gg \hbar$) or Lorentzian ($\Sigma\tau \ll \hbar$) linewidth dominates. ΔV is then extracted by approximating a Gaussian ($\Delta V_G = 2\sqrt{2ln2}\Sigma$) or Lorentzian ($\Delta V_L = 2\Sigma^2\tau/\hbar$) emission lineshape.

The contribution of the Gaussian ($\Delta V_G = 2\sqrt{2ln2}\Delta$) and Lorentzian ($\Delta V_L = 2\Gamma$) components to the overall linewidth can be extracted directly from the fitted function $C_2(t)$. The convolution of a Lorentzian with a Gaussian lineshape results in an overall spectral lineshape with a Voigt profile that has a linewidth that may be approximated by $\Delta V_V \approx 0.5346\Delta V_L + \sqrt{0.2166\Delta V_L^2 + \Delta V_G^2}$.

For the four QDs shown in Fig. 6.5a–d the values for overall linewidth given by $C_1(t)$ and $C_2(t)$ are displayed in Table 6.1. It is observed that the values obtained from the two models agree well with one another. Table 6.1 also contains the Gaussian and

Lorentzian contributions to the overall linewidths for the four InP QDs displayed in Fig. 6.5a–d, as extracted by the function $C_2(t)$.

If the lineshape of the emission is known a coherence time, T_2, can be evaluated from the linewidths. For a Lorentzian lineshape the coherence time is given by $T_{2L} = 2\hbar/\Delta V$ and for a Gaussian $T_{2G} = \hbar/\Delta V$. The coherence time for a Voigt lineshape T_{2V} is known to lie between the two values T_{2L} and T_{2G} as it is a convolution of Lorentzian and Gaussian contributions.

QD_C has the longest estimated T_{2V} of between $T_{2L} = 16.5$ ps and $T_{2G} = 33$ ps. When evaluated alongside the typical lifetime of our InP QDs, we extract an indistinguishability ratio, $R = \frac{T_2}{2t_1} = 0.006$.

If we consider the shortest recorded T_1 time for a Purcell enhanced InP QD, where $T_1 = 0.09$ ns the notional indistinguishability ratio remains low, $R = 0.14$. Without improving the T_2^* time for InP QDs, an approximate tenfold improvement in the current maximum Purcell enhancement is required to achieve a $2T_1 = T_2^*$ and $R = 0.5$.

The typical linewidths for QDs in our sample agree with reported linewidths for InP QDs, [23, 25, 26] where T_2's of approximately 10 ps or less were reported. The coherence time for InP QDs are an order of magnitude shorter than for InAs QDs whilst their typical lifetimes remain comparable [22]. This significantly increases the relative difficulty in producing indistinguishable single photon sources using InP QDs compared with InAs QDs.

The understanding of underlying processes dominating the decoherence of the quantum dot emission is an important step toward improving the coherence times for this material system. By distinguishing between two contributions to the decoherence process, the function $C_2(t)$, allows us to extract information about the cause of the short coherence times.

6.8 Photonic Crystal Cavity Mode Detuning

Figure 6.6a plots the linewidths of the voigt lineshape found by the convolution of the Gaussian and Lorentzian components extracted from the fitted function $C_2(t)$,

Table 6.1 Summary of the overall linewidths, ΔV, from the two models $C_1(t)$ and $C_2(t)$ for four InP QDs

QD	$C_2(t)$ (μeV)			$C_1(t)$ (μeV)
	ΔV_L	ΔV_G	ΔV_V	ΔV
A	95 ± 17	65 ± 19	129	123
B	13 ± 4	61 ± 3	68	77
C	23 ± 3	25 ± 4	39	28
Bulk	61 ± 9	42 ± 10	83	78

The Lorentzian ΔV_L and Gaussian ΔV_G contribution to the overall linewidth are also displayed for $C_2(t)$

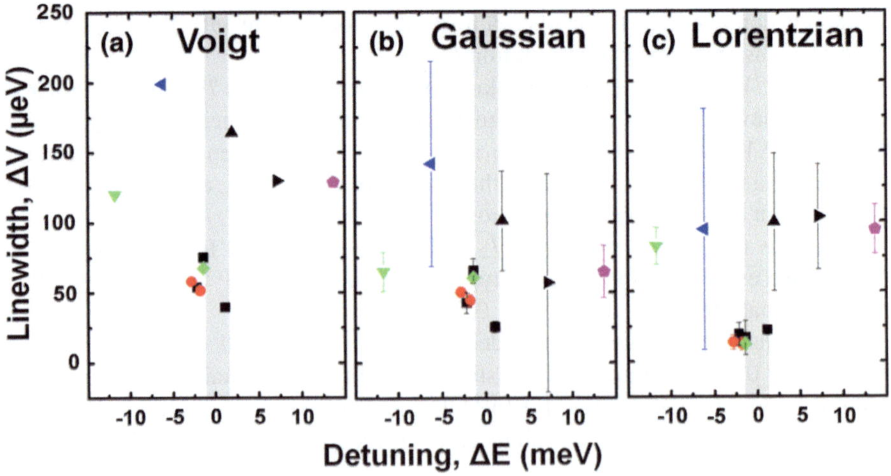

Fig. 6.6 Plots of the **a** Voigt **b** Gaussian **c** Lorentzian contribution to QD linewidths, ΔV, extracted from $C_2(t)$ as a function of detuning from the nearest photonic crystal mode, ΔE. The *grey* region represents a typical FWHM for the photonic crystal modes studied with $Q \sim 700$. Each individual QD transition is represented by one of eight unique shapes. Where a single shape is repeated, data has been recorded for the same QD transition on separate occasions when the relative QD to mode detuning has been tuned through nitrogen condensation. The data points are also colour coded (*pink, red, green, black, blue*) to group QD transitions that were observed in the same photonic crystal cavity from the five cavities studied

for several more QDs that were studied. The linewidths are plotted as a function of the detuning from the closest photonic crystal cavity mode, ΔE. The data is colour coded to represent the cavity in which they were observed and shape coded for individual QD transitions. Each individual QD transition is represented by one of eight unique shapes. Where a single shape is repeated, data has been recorded for the same QD transition on separate occasions when the relative QD to mode detuning has been tuned through nitrogen condensation. The data points are also colour coded (pink, red, green, black, blue) to group QD transitions that were observed in the same photonic crystal cavity from the five cavities studied [27].

In the cavities studied, the modes were broad, with a relatively low $Q \sim 700$. The grey shaded region on the plot indicates the average FWHM of the modes which is approximately 2.5 meV.

The data in Fig. 6.6a suggests that two regimes of QD linewidth exist depending on the relative mode to QD detuning, ΔE. The QD transitions that are close to resonance with a cavity mode, $\Delta E < 3$ meV, exhibit a much reduced linewidth, $\Delta V =$ 28–75 μeV than far off resonant transitions, $\Delta E > 3$ meV, where $\Delta V = 120$–200 μeV. This suggests the cavity mode may influence the coherence time of photons emitted by the QD.

A Purcell enhancement of the radiative lifetime can be ruled out as a cause for the large increase in T_2 for on resonant QDs because the T_2 time is still dominated by T_2^*, making any reduction in T_1 insignificant.

Figure 6.6b displays the Gaussian contribution to linewidth as a function of the detuning, ΔE, from the closest photonic crystal mode where $\Delta E = E_{mode} - E_{QD}$. It shows an even distribution of the Gaussian contributions irrespective of ΔE, with an average contribution of 65.5 µeV.

The function $C_2(t)$ associates the Gaussian component of linewidth to fluctuations caused by trapped charges. The even distribution of the Gaussian contributions irrespective of ΔE indicates that the interactions between the QDs and trapped charges are unaffected by resonances betweeen a QD transition and the photonic crystal modes, and the Gaussian contribution to linewidth remains similar to the Gaussian contribution to linewidth of the QD transtion observed in the bulk. The variation that is observed in the Gaussian component is likely to be a consequence of the variation in proximity of the QDs to trapping sites.

The large error bars that exist on some data points are due to the two related terms in $C_2(t)$. When the relative weighting of Gaussian to Lorentzian contributions become comparable the magnitude of the error bars increase. We also observe larger error bars when the overall coherence length of the QD line is low because there are fewer available data points to use in the fit.

Figure 6.6c displays the Lorenztian contribution to the overall linewidth, V_L, as a function of detuning, ΔE. Unlike the data in Fig. 6.6b there is a significant difference between near resonant ($\Delta V_L \sim$ 15 µeV) and far off resonant ($\Delta V_L \sim$ 90 µeV) QD transitions. The function $C_2(t)$ associates the Lorentzian component of linewidth to fluctuations caused by acoustic phonons, so the variation in Lorentzian component indicates that QDs near resonance have a significantly reduced phonon interaction compared with far off resonant QDs.

Although this behaviour is not well understood, evidence of linewidth broadening for off resonant QDs has recently been observed in InAs QDs where it was attributed to coupling with the continuum of states in the wetting layer and/or neighbouring GaAs [28]. There has also been much interest in the off resonant feeding of photonic crystal modes and it has been shown that this process is mediated by acoustic phonons for near resonant QDs but does not occur for far off resonant QD lines due to a significant change in the acoustic phonon coupling strength between near resonant and far off resonant QD transitions [17, 29, 30].

We have seen in Fig. 6.6 that the Gaussian contribution to the linewidth is generally the limiting factor for QDs with the longest observed coherence times, i.e., those resonant with photonic crystal modes. Since the Gaussian contribution recorded for our dots is 65.5 µeV, this limits the mean coherence time to around 10 ps.

6.8.1 Discussion

The data shown in Fig. 6.6 that displays the variation in Lorentzian and Gaussian contributions to the linewdith of a QD transition in relation to the detuning of the transition to a cavity mode and the subsequent interpretation based on acoustic phonons and trapped carriers is not conclusive. Questions remain about the origin of the decoherence process that take place within InP QDs and the effect photonic crystals and mode resonances have in this.

It is clear that from the data presented in Fig. 6.6 that there are variations in the coherence times of the QD transitions, and a relationship between the linewidth and the detuning of the photonic crystal mode is possible.

To confirm that there is a relationship between the linewidth of the QD transition and the detuning of the photonic crystal mode it would be desirable to observe a decrease in the overall linewidth of a QD transition as it is tuned into resonance with a cavity mode. To support the interpretation that any decrease is in some way related to the acoustic phonon interaction it would be desirable to observe, in particular, a decrease in the Lorentzian linewidth component as the QD transition is tuned into resonance with a cavity mode and investigate whether the interaction energy between phonons and the QD transition is large enough to account for the change in linewidth.

Attempts were made to record data from the same QD transition as the photonic crystal cavity mode was tuned, and these are displayed in Fig. 6.6, however, no data was recorded that conclusively supported the hypothesis that the linewidth would decrease on resonance. No systematic method for tuning the cavity mode was available at this time and temperature tuning of a nearby QD into resonance was discarded as a means by which to achieve this systematic tuning as the induced temperature changes would also induce a linewidth variation in the QD transition that could not be separated from the resonance affect.

At this stage of the investigation preparations were being made to continue with tuning the PCC mode through nitrogen condensation. Part of the preparations included cleaning the sample to recover higher Q-factor so that a smaller range of tuning would be required. Unfortunately, cleaning the samples altered significantly both the Q-factor and energy of the PCC modes, so that they were spectrally located in a high density part of the QD ensemble. The high Q-factors also made it difficult to filter QD emission from PCC mode emission when resonant. As a result, no comparison between the on resonance and off resonance linewidths of a single transition exist.

After suggesting that phonon mediated feeding of the photonic crystal mode may affect the QD transition linewidth when near-resonant with a cavity mode, it is interesting to consider further the data displayed in Fig. 6.6. We may expect to observe an asymmetry in the linewidths between the QDs detuned to high energy side compared to the low energy side of the cavity mode, due a difference in efficiency of photon absorption versus emission [30]. Although the lack of data for low energy near-resonant QD transitions make it difficult to draw any strong conclusions the data in Fig. 6.6c suggests that an asymmetry may be present. The data point closest to the FWHM of the cavity mode on the low energy side exhibits a high Lorentzian component of $\sim 100\,\mu\text{eV}$ whilst the QD transitions detuned by the same magnitude on the high energy side have a low Lorentzian component of $\sim 20\,\mu\text{eV}$.

6.9 Summary

InP QDs suffer from short coherence times relative to InAs QDs and this is restricting their development as a source of indistinguishable single photons. By means of Fourier transform spectroscopy we have been able to analyse the dephasing processes within the InP QDs. Although we have seen that resonances with photonic crystal modes may reduce the Lorentzian component of the QD transition linewidth which is the contribution attributed to dephasing caused by fast fluctuations in the QD environment due to acoustic phonons, we have found that slow modulations attributed to trapping of carriers still limit the coherence time and indistinguishability ratios that may be achieved for all InP QDs. The realisation of an indistinguishable source of single photons from an InP based QD system will benefit significantly from a reduction in the dephasing caused by trapped carrier processes and this will most likely be achieved by an improvement in the material quality.

References

1. J. Rarity, P. Tapster, Ground to satellite secure key exchange using quantum cryptography. New Journal of Phys. **82** (2002)
2. W.-M. Schulz, R. Roßbach, M. Reischle, G. Beirne, M. Bommer, M. Jetter, P. Michler, Optical and structural properties of InP quantum dots embedded in $(Al_xGa_{1\hat{a}\acute{L}\check{S}x})_{0.51}In_0.49P$. Phys. Rev. B **79**, 1–8 (2009)
3. M. Barth, N. Nüsse, B. Löchel, O. Benson, Controlled coupling of a single-diamond nanocrystal to a photonic crystal cavity. Opt. Lett. **34**, 1108–1110 (2009)
4. M. Barth, J. Kouba, J. Stingl, B. Löchel, O. Benson, Modification of visible spontaneous emission with silicon nitride photonic crystal nanocavities. Opt. Express **15**, 17231–17240 (2007)
5. M. Barth, N. Nüsse, J. Stingl, B. Löchel, O. Benson, Emission properties of high-Q silicon nitride photonic crystal heterostructure cavities. Appl. Phys. Lett. **93**(2), 021112 (2008)
6. K. Rivoire, A. Faraon, J. Vuckovic, Gallium phosphide photonic crystal nanocavities in the visible. Appl. Phys. Lett. **93**(6), 063103 (2008)
7. Z. Zhang, T. Yoshie, X. Zhu, J. Xu, A. Scherer, Visible two-dimensional photonic crystal slab laser. Appl. Phys. Lett. **89**(7), 071102 (2006)
8. I.J. Luxmoore, E.D. Ahmadi, N.A. Wasley, A.M. Fox, A.I. Tartakovskii, A.B. Krysa, M.S. Skolnick, Control of spontaneous emission from InP single quantum dots in GaInP photonic crystal nanocavities. Appl. Phys. Lett. **97**(18), 181104 (2010)
9. W.-H. Chang, W.-Y. Chen, H.-S. Chang, T.-P. Hsieh, J.-I. Chyi, T.-M. Hsu, Efficient single-photon sources based on low-density quantum dots in photonic-crystal nanocavities. Phys. Rev. Lett. **96**, 3–6 (2006)
10. D.G. Gevaux, P. Atkinson, D. Anderson, A.J. Bennett, S.P. Bremner, J. Griffiths, D.J.P. Ellis, R.M. Stevenson, G.A.C. Jones, D.A. Ritchie, A.J. Shields, Controlling spontaneous emission from quantum dots using photonic crystal microcavities. Phys. Status Solidi (C) **3**, 3676–3679 (2006)
11. D. Englund, D. Fattal, E. Waks, G. Solomon, B. Zhang, T. Nakaoka, Y. Arakawa, Y. Yamamoto, J. Vučković, Controlling the spontaneous emission rate of single quantum dots in a two-dimensional photonic crystal. Phys. Rev. Lett. **95**, 2–5 (2005)
12. H.J. Kimble, The quantum internet. Nature **453**, 1023–1030 (2008)

13. G. Khitrova, H. Gibbs, M. Kira, S. Koch, A. Scherer, Vacuum Rabi splitting in semiconductors. Nat. Phys. **2**(2), 81 (2006)
14. T. Yoshie, A. Scherer, J. Hendrickson, G. Khitrova, H.M. Gibbs, G. Rupper, C. Ell, O.B. Shchekin, D.G. Deppe, Vacuum rabi splitting with a single quantum dot in a photonic crystal nanocavity. Nature **432**, 200–203 (2004)
15. Y. Akahane, T. Asano, B. Song, S. Noda, High-Q photonic nanocavity in a two-dimensional photonic crystal. Nature **425**(October), 944–947 (2003)
16. M. Kaniber, A. Kress, A. Laucht, M. Bichler, R. Meyer, M.-C. Amann, J.J. Finley, Efficient spatial redistribution of quantum dot spontaneous emission from two-dimensional photonic crystals. Appl. Phys. Lett. **91**(6), 061106 (2007)
17. M. Kaniber, A. Laucht, A. Neumann, J. Villas-Bôas, M. Bichler, M.-C. Amann, J. Finley, Investigation of the nonresonant dot-cavity coupling in two-dimensional photonic crystal nanocavities. Phys. Rev. B **77**(16), 1–4 (2008)
18. J. Gérard, B. Sermage, B. Gayral, B. Legrand, E. Costard, V. Thierry-Mieg, Enhanced spontaneous emission by quantum boxes in a monolithic optical microcavity. Phys. Rev. Lett. **81**, 1110–1113 (1998)
19. E. Knill, R. Laflamme, G.J. Milburn, A scheme for efficient quantum computation with linear optics. Nature **409**, 46–52 (2001)
20. S. Laurent, S. Varoutsis, L. Le Gratiet, A. Lemaître, I. Sagnes, F. Raineri, A. Levenson, I. Robert-Philip, I. Abram, Indistinguishable single photons from a single-quantum dot in a two-dimensional photonic crystal cavity. Appl. Phys. Lett. **87**(16), 163107 (2005)
21. C. Santori, D. Fattal, J. Vucković, G.S. Solomon, Y. Yamamoto, Indistinguishable photons from a single-photon device. Nature. **419**, 594–597 (2002)
22. A. Berthelot, I. Favero, G. Cassabois, C. Voisin, C. Delalande, P. Roussignol, R. Ferreira, J.M. Gérard, Unconventional motional narrowing in the optical spectrum of a semiconductor quantum dot. Nat. Phys. **2**, 759–764 (2006)
23. K. Bando, Y. Masumoto, Narrowing of exciton linewidth of a quantum dot with increasing temperature. Phys. Status Solidi (C) **6**, 57–60 (2009)
24. P. Blome, M. Wenderoth, M. Hübner, R. Ulbrich, J. Porsche, F. Scholz, Temperature-dependent linewidth of single InP/$Ga_xIn_{1-x}P$ quantum dots: interaction with surrounding charge configurations. Phys. Rev. B **61**, 8382–8387 (2000)
25. T. Aichele, V. Zwiller, O. Benson, Visible single-photon generation from semiconductor quantum dots. New J. Phys. **6**, 90–90 (2004)
26. V. Zwiller, T. Aichele, O. Benson, Single-photon Fourier spectroscopy of excitons and biexcitons in single quantum dots. Phys. Rev. B **69**, 1–4 (2004)
27. S. Mosor, J. Hendrickson, B.C. Richards, J. Sweet, G. Khitrova, H.M. Gibbs, T. Yoshie, A. Scherer, O.B. Shchekin, D.G. Deppe, Scanning a photonic crystal slab nanocavity by condensation of xenon. Appl. Phys. Lett. **87**(14), 141105 (2005)
28. A. Majumdar, A. Faraon, E. Kim, D. Englund, H. Kim, P. Petroff, J. Vučković, Linewidth broadening of a quantum dot coupled to an off-resonant cavity. Phys. Rev. B **82**, 1–5 (2010)
29. Y. Ota, M. Shirane, M. Nomura, N. Kumagai, S. Ishida, S. Iwamoto, S. Yorozu, Y. Arakawa, Vacuum rabi splitting with a single quantum dot embedded in a H1 photonic crystal nanocavity. Appl. Phys. Lett. **94**(3), 033102 (2009)
30. M. Winger, T. Volz, G. Tarel, S. Portolan, A. Badolato, K. Hennessy, E. Hu, A. Beveratos, J. Finley, V. Savona, A. Imamoğlu, Explanation of photon correlations in the far-off-resonance optical emission from a quantum-dot-cavity system. Phys. Rev. Lett. **103**, 11–14 (2009)

Chapter 7
Development of Additional Technological Approaches

7.1 Introduction

The development of quantum optical circuits provides an opportunity to study novel systems and quantum phenomena as yet not fully understood. However, the development of functional and controllable systems requires a range of technology to facilitate the integration of a large, and ever increasing, number of components. The purpose of the technology is to produce circuits that are inherently forgiving to the limitations of the fabrication that is currently available by providing a degree of tunability and control to a number of aspects in these systems.

Typically this includes tuning the energy of cavity or waveguide modes and QD transitions as well as reducing the reliance on chance that is inherent in fabricating photonic structures that include self-assembled QDs.

The technology is often well understood but difficult to integrate with on-chip circuits. Hence there is a need for constant development of techniques to provide new and improved capabilities to the circuits. In turn, this furthers the progress in the fundamental understanding of these systems.

In most cases the development of a technology that provides a new capability may not be a long term solution but will provide some progress and new understanding of the quantum optical systems. Therefore, I feel it appropriate and necessary to detail a number of technologies we have been developing for use in the quantum-optical systems.

7.2 Quantum Dot Registration

To observe the strong modifications to quantum dot emission that photonic crystal cavities can provide, the location of the quantum dot within the cavity is critical. The strength of coupling (g) between the QD and cavity depends on the strength of electric field $\vec{E}$ at the location of the QD

N. A. Wasley, *Nano-photonics in III–V Semiconductors for Integrated Quantum Optical Circuits*, Springer Theses, DOI: 10.1007/978-3-319-01514-9_7,

$$g_{(\vec{r})} = \frac{1}{\hbar}|\vec{\mu}_{12}.\vec{E}_{(\vec{r})}| \tag{7.1}$$

where $\vec{E}$ is a function of position $\vec{r}$.

The strength of the electric field across the photonic crystal cavity is not uniform. The standing wave of the electric field mode contains nodes and antinodes. The coupling between the QD and cavity is strongest when the area of highest electric field, the antinode, and the QD are spatially resonant.

Due to the random nucleation location of QDs in the self assembled growth process, schemes for fabricating coupled QD-cavity systems have generally relied on chance, with the fabrication and measurement of many systems required to find a QD located at the antinode of the electric field. The lack of a suitable technique is a prohibiting factor in the development of solid state quantum emitter devices as well as the fundamental investigations of cavity QED.

This section describes a deterministic method of fabrication that registers the position of a single quantum dot of given spectral characteristics to an accuracy of a few tens of nanometres with the intention of patterning an optical device around the QD.

7.2.1 Required Positional Accuracy

The required accuracy for QD registration and photonic crystal positioning depends on the size of the antinodes in the cavity modes to be exploited. The size of the antinodes depends on the material system, cavity design and cavity mode being studied, but typically an accuracy of 50–100 nm is required.

The electric field profile calculated by FDTD analysis of a fundamental mode in a L3 type cavity with a lattice period $a = 187$ nm and hole radius $r = 0.3a$ is shown in Fig. 7.1. The central antinode can be seen as the red dot with an approximate spatial diameter of 50 nm.

7.2.2 Registration Technique

There have been a number of attempts to develop such a technique, that include the use of AFM to detect small hills on the surface of a sample caused by the presence of QDs in the quantum dot layer [1] and registration of QDs with in-situ photolithography [2–4]. These techniques although successful are still significantly experimentally challenging.

In 2009 an all optical method for measuring the position of individual QDs that deterministically achieved strong coupling between the QD and photonic crystal cavity requiring no in-situ photolithography due to the use of pre-fabricated markers was

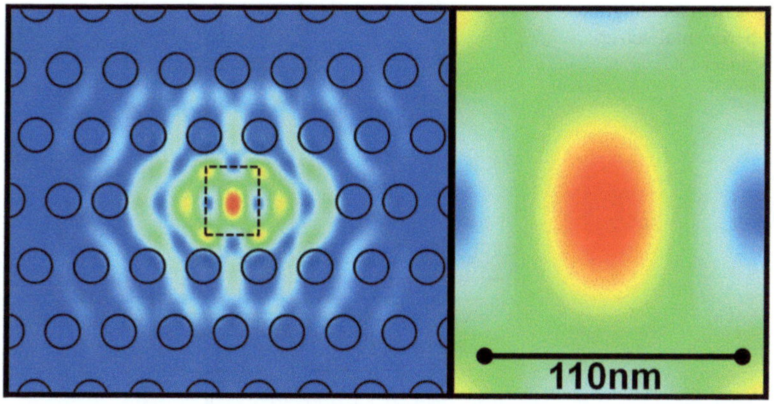

Fig. 7.1 The electric field profile for the fundamental mode of an L3 type cavity with a lattice period $a = 187$ nm, hole radius $r = 0.3a$ and end hole shift $s = 0.18$ calculated with fintie difference time domain (FDTD) analysis. A *Blue–Red scale* is used such that red colours represent areas of high electric field or antinodes, and the blue colours are of low field, nodes. The central antinode can be seen as the red dot with an approximate spatial diameter of 50 nm

developed [5]. The registration of the quantum dot relies on the precise determination of the optical centre of emission and takes place in the measurement set-up.

Based on this significant simplification in the methodology, we develop a similar technique but differing in a number of aspects, including material system, marker identification and QD emission detection.

Gold open box markers are fabricated on an InP QD sample using electron beam lithography and metal evaporation. InP QDs are used as they provide greater PL signal compared to InAs QDs, due to the increased efficiencies of Si based detector systems to photons at wavelengths of InP QD emission around 670 nm compared with InAs QDs at 970 nm.

μPL with excitation at 633 nm is used to locate a suitable QD within a set of the open box markers. The spectrometer is set at the peak emission wavelength for the identified QD and a closed loop piezo-stage performs multiple scans in perpendicular x and y directions over the gold markers and the QD.

Figure 7.2, displays a schematic representation of the experimental set-up used to register the QDs.

The QD emission and HeNe laser reflection is collected with an objective lens before being split with a 90:10 beam splitter. 10 % of the collected light is heavily attenuated before the signal is recorded directly by an avalanche photo diode (APD) without any spectral filtering. The vast majority of signal returned by this APD is due to the reflected incident HeNe laser light, and is used to identify the location of the gold markers. When the laser scans over the gold markers, there is a peak in the signal as they are highly reflective compared to the semiconductor surface.

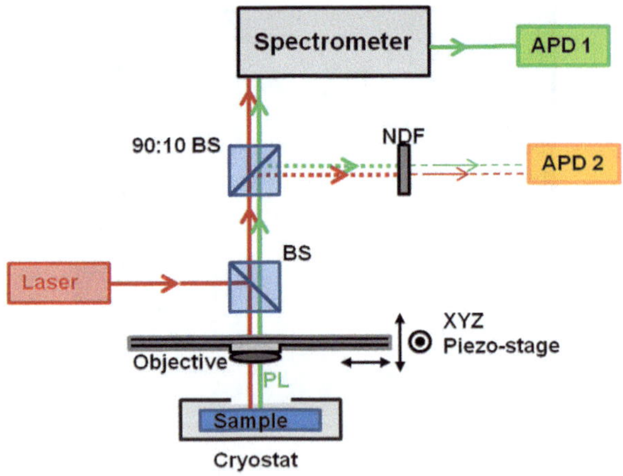

Fig. 7.2 Schematic diagram of the experimental set-up used to register QDs

The remainder of the collected light is spectrally filtered by the spectrometer which is used to isolate the wavelength corresponding to the initially identified QD before the signal is recorded by a second APD.

Although the wavelength of the scanning laser is larger than the required positioning accuracy, the positioning method locates the centre of emission pattern by fitting the emission peak and averaging over many scans. The measured position of the QD and markers are approximately normally distributed. With thirty scans the location of the dot can be identified to a statistical error of around 5 nm.

7.2.3 Two-channel Mapping

Once a suitable QD has been found within a set of markers through a manual search of the area, a two channel map of the marker set is taken. This serves as a visual reference for the location of the QD and is used to find the rotation angle of the marker axis relative to the piezostage axis.

The mapping takes the form of a raster scan; the piezostage steps along one axis, before returning to the initial position, taking one step in the perpendicular axis and repeating until the area of the marker set is covered. A measurement on both APDs is taken simultaneously for each step. In this way, a map of both the markers and QD can be created. An example of this two channel map is shown in Fig. 7.3.

The target QD is clearly visible in Fig. 7.3b. It is also possible to make out the outline of the marker set, due to the reduction in background counts from QD emission under the markers.

Figure 7.3a is the map created from the laser reflections from the gold markers. The outline of the open box markers are clearly visible. It is this map that is used to

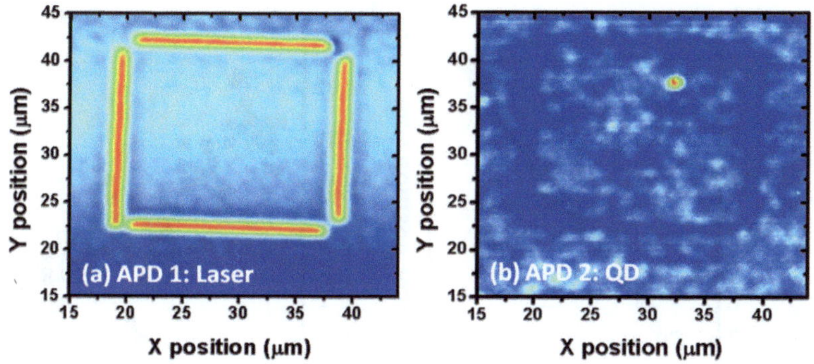

Fig. 7.3 **a** Intensity map of the 633 nm laser reflections from of a marker set with 120 × 120 data points. **b** PL intensity map of QD emission from the same marker set with 120 × 120 data points

calculate the angle of rotation between the piezostage axis and the marker set, which is critical if the position on the QD relative to the markers is to be extracted.

7.2.4 Extracting the Rotation Angle

A rotation angle, θ, exists between the axis of the piezoelectric stage and that defined by the straight lines of the marker set. By extracting the position of a marker on the piezostages from Gaussian fits to the spatial reflection peaks, the co-ordinates of the marker position can then be plotted.

The gradient of a linear fit, m, to the line of marker positions then provides the rotation angle between the axis of piezostage movement and the marker axis, θ, such that $90 - tan\theta = 1/m$. A plot of the marker positions with a linear fit can be seen in Fig. 7.4. From this a rotation angle of $\theta = 1.927°$ is extracted.

7.2.5 System Drift

The stability of the system is an important factor in determining whether an accurate position for the QD relative to the fabricated markers can be extracted. If the sample drifts relative to the focus of the objective lens on a time scale comparable with the time required for a line scan to cross the marker and move to the QD, correction factors or improvements to the stability of the system would be required.

By monitoring the position of a marker as recorded by the piezostage for 100 line scans and plotting the results against time as can be seen in Fig. 7.4b. A linear fit gives an estimate of the drift as 1.4 nm/min. A typical line scan takes approximately 30 s, with the time for the stage to move between the marker and the dot being significantly

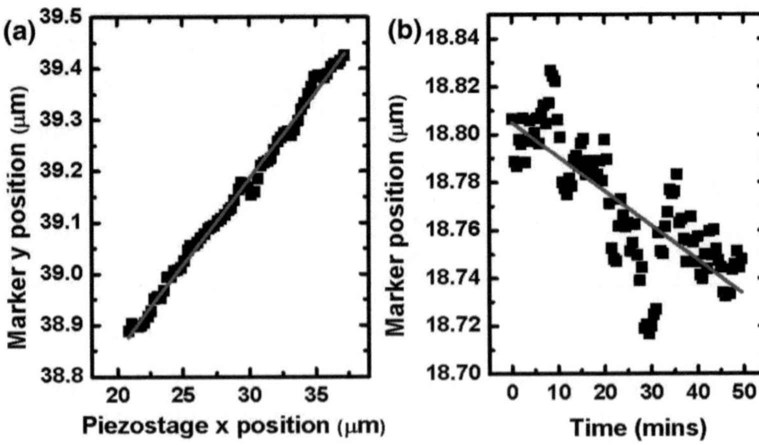

Fig. 7.4 **a** Linear fit, $y = 0.03364x + 38.17$, to the marker position data extracted from the laser reflection map in Fig. 7.3. **b** The position of a marker relative to the piezostage as a function of time. A linear fit, $y = -0.00144x + 18.805$, estimates the drift of the set up to be approximately 1.4 nm/min

less. Therefore the system drift may induce some error in the positioning of the system but this is likely to be less than 1 nm for any single scan, an acceptable level and comparable to the system drift recorded elsewhere [2].

7.2.6 Line Scans

After the two channel mapping has taken place, multiple line scans across the dot and markers are completed in orthogonal orientations to produce an accurate final co-ordinate.

Before the line scans take place the positioning of the system is optimised by raster scanning the area close to the QD. The step size of the line scan is reduced to 150 nm from the 250 nm steps used in the mapping process in order to improve the data density for the peak fitting. This value is time limited rather than the resolution of the stage.

Gaussian fits to the QD peak and both marker peaks from the two channel line scan data in both orientations are then made. The peak position determines the recorded position of the relevant feature.

Figure 7.5a and b are sections of the lines scans for a marker set. Figure 7.5a is the QD emission and Fig. 7.5b is the reflected laser signal from one marker.

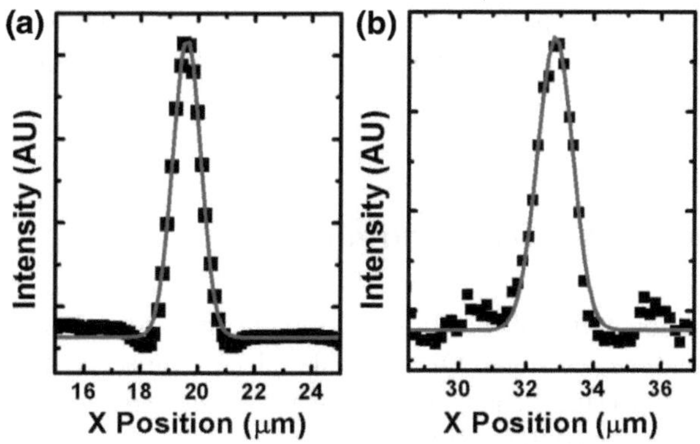

Fig. 7.5 Sections of line scans that contain **a** quantum dot emission peak and **b** laser reflection peak from a marker. Both are fitted with Gaussian curves to extract the position of peak intensity

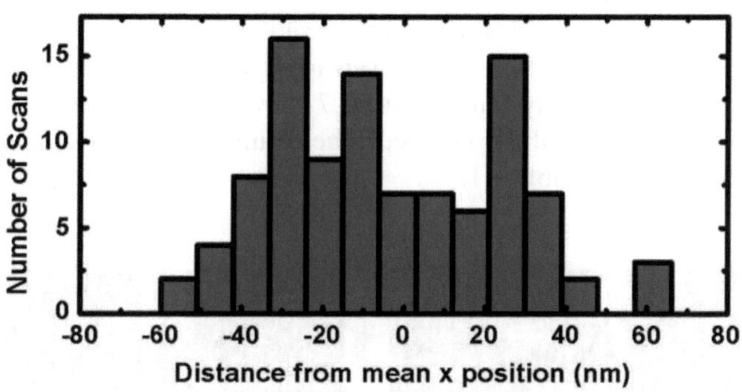

Fig. 7.6 Histogram of quantum dot position data relative to the mean distance, $\bar{x}$, for all data from a marker. Standard error of the mean for this approximately normal distribution of 100 data points is $SE_{\bar{x}} = 2.6\,\text{nm}$

7.2.7 Distribution of Recorded QD Positions

For each line scan the QD position relative to a marker is extracted by subtracting the position of the QD from the position of the marker. The distribution of this relative positioning for the set of line scans is assumed to be approximately normally distributed. From this distribution a mean value, $\bar{x}$, and a standard deviation from the mean, σ, is extracted. This is used to provide the standard error of the mean $SE_{\bar{x}} = \sigma/\sqrt{n}$, where n is the number of line scans. In this way, the position of the QD can be located more accurately than the wavelength of the incident laser.

The histogram of the QD position relative to the mean QD location for a registration run with $n = 100$ is shown in Fig. 7.6. The standard error for the distribution is approximately 2.6 nm. In comparison Thon presented a similar standard error of 2 nm.

Once this process is completed for the orthogonal axis and the result corrected for the angle of rotation. A co-ordinate for the position of the QD relative to the pre-fabricated markers is extracted.

7.2.8 E-beam Fabrication Accuracy

Once the registration of the QD position is complete, the fabrication of photonic crystal cavities around the registered QDs can take place. The accuracy of the photonic crystal fabrication is a significant and non trivial step in the deterministic positioning of a QD at the antinode of the cavity mode.

To test the positional accuracy of the electron beam lithography machine used to write the photonic crystal pattern, marker sets were pattered on to a semiconductor wafer in a similar process to the fabrication of the open box markers. The E-beam machine patterns a set of identical markers aligned to the first set. By measuring the difference in the alignment of the marker sets using a SEM, we are able to estimate the accuracy of the E-beam system. Figure 7.7 shows an SEM image of the aligned markers with the spatial analysis labelled. The results for a vertical and horizontal marker test are shown in Table 7.1.

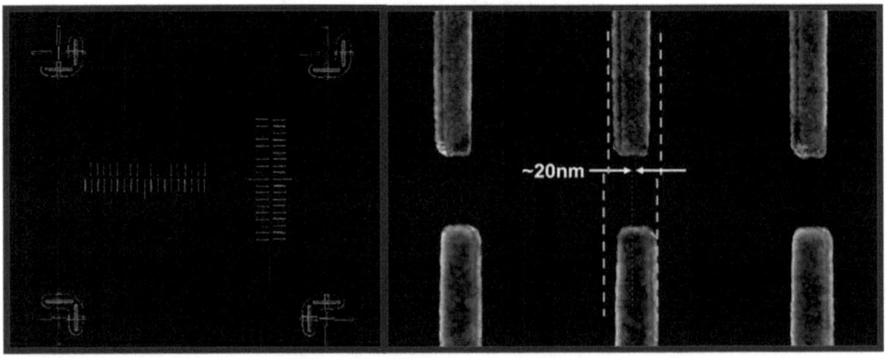

Fig. 7.7 SEM image of the alignment test markers to determine the accuracy of the electron beam lithography. Alignment guides and approximate offset is labelled. The shadowing is caused by a double evaporation process

Table 7.1 Electron-beam lithography alignment results

Mark set	Horizontal misalignment (nm)	Vertical misalignment (nm)
1	6.5 ± 2.9	9.5 ± 3.8
2	4.5 ± 1.9	20.2 ± 5.1
3	2.4 ± 2.4	16.1 ± 2.4

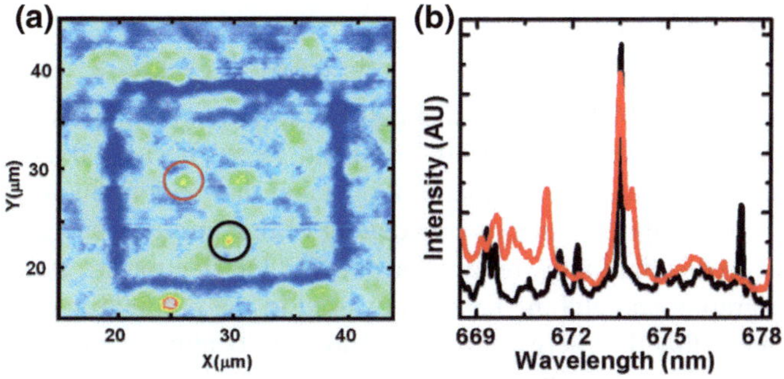

Fig. 7.8 **a** A PL map of one set of markers with the spectral window set at ~973.5 nm. Two spectra taken at the locations circled in (**a**) are colour coded and displayed in (**b**), **b** QD emission lines with wavelengths of 973.522 and 973.518 nm are observed

The E-beam lithography was found to have a typical positional accuracy of around 10–20 nm, an accuracy level that confirms the E-beam lithography available is capable of writing the photonic crystals to the required accuracy for deterministically placing quantum dots at the antinode of photonic crystal cavities.

7.2.9 Future Work

The final stage in producing a complete system where the QD has been deterministically placed within a photonic crystal cavity has not yet been completed due to operational difficulties with the electron beam lithography machine at the time this project was being under taken. This remains the final validation process to take place before we consider the registration system to be fully operational. Given that this process is eventually successfully completed a number of interesting research areas become available.

In wafers with a high QD density there is a chance when scanning the markers for a specific QD, another QD emission line will be in resonance, such that the emission peaks are less than the linewidth apart. This creates the possibility of developing deterministically coupled quantum dots linked with photonic crystal or plasmonic waveguides [6].

The realistic possibility of attaining this is demonstrated in the system presented in Fig. 7.8. It displays a PL map of one set of markers with the spectral window set at 973.5 nm. Due to the high dot density in this wafer region a number of features can be observed but two locations of particularly high signal can be identified within the marker system. The two spectra taken at the location of each of these signals, as circled in Fig. 7.8a are colour coded and displayed in Fig. 7.8b, QD emission lines with wavelengths of 973.522 and 973.518 nm can be identified, suitable for use in a photonic structure that couples the two QDs.

7.3 Tuning of QD Emission

It is not currently possible to deterministically fabricate a system in which the cavity mode and a QD transition are spectrally resonant. This is because fluctuations in the size and shape of self assembled quantum dots and the random nucleation location as they form cause inhomogeneous broadening of the emission energy and make it impossible to predict the energy or location of a QD. Fabrication errors inherent in the E-beam lithography and etching techniques used in the production of the photonic crystal also mean that the energy of a cavity mode or location of a photonic structure cannot be exactly predetermined.

Even if better control of QD emission energy and location as well as improved fabrication were achieved it is still desirable to be able to control spectral detuning between a QD and a cavity mode. Hence there is a need for a method to modify the system such that a QD emission line or cavity mode may be tuned into and away from resonance.

There have been a number of techniques developed to provide this type of control over the system, including wet chemical digital etching [7], thin film gas deposition [8] and AFM oxidisation [9].

The alternative to tuning the cavity mode emission is to tune the QD emission. Temperature tuning is a widely used technique for tuning a QD into resonance with a cavity mode. The QD bandgap and photonic crystal mode energy are both temperature dependent. The QD bandgap has a greater dependence on temperature than the photonic crystal mode and so by controlling the temperature of the sample the QD energy may be blue shifted at a greater rate than the mode. This is easy to achieve and has been shown to work on a localised cavity mode [10]. However, there is only a small tuning range before carriers are lost and there is a degradation in the quality of recorded spectra as the PL intensity from individual QD transitions is reduced.

Another technique for tuning the QD emission is to use the *quantum confined Stark effect*. By applying an electric field across a QD the bandgap and potential well are modified. This causes the wavefunctions to separate in space and the single particle energy levels within the QD to lower. Although the increase in spatial separation also increases the radiative recombination lifetime of an electron-hole pair, it is possible to tune the QD emission energy through the magnitude of the applied field before this significantly affects the emission intensity.

In this section a P-I-N diode structure is used to tune QD emission by varying an external bias. The importance of an electrical tuning technique is that it readily extends to on-chip devices and has the potential for high frequency modulation [11].

7.3.1 P-I-N Diode Device Structure

GaAs based P-I-N diode structures that incorporate an InAs QD layer within the intrinsic GaAs region, located between p-type and n-type GaAs layers, were fabricated using photolithography and a wet chemical etch, as described in Chap. 2,

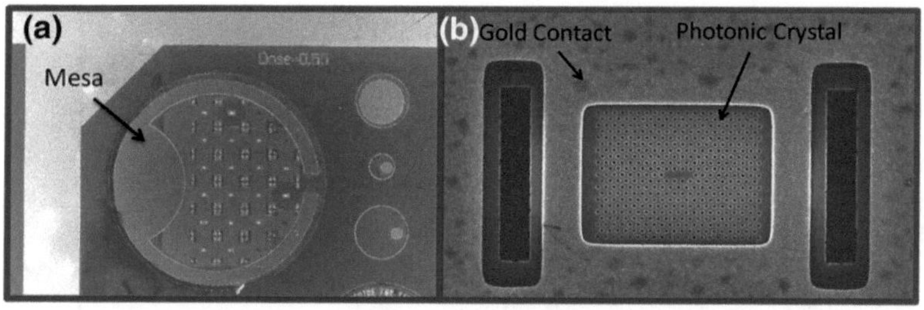

Fig. 7.9 SEM images of the photonic crystal P-I-N diodes. **a** A complete diode mesa. **b** A photonic crystal situated on the mesa labelled in (**a**) shown with the gold contact aperture around the photonic crystal

Sect. 2.2. The etch is used to create a mesa that exposes the underlying n-type GaAs, the round mesas can be seen in Fig. 7.9a. E-beam lithography and $SiCl_4$ reactive ion etching (RIE) were used to pattern L3 type photonic crystals through the exposed diode before a HF etch of the sacrificial AlGaAs layer creates an air bridge under the photonic crystal.

To complete the structure, a metal evaporation onto the n-type and p-type GaAs layers provide electrical contacts to the diode. The n-type contact is an annealed In-Ge-Au contact to ensure an ohmic contact with the device, while the p-type contact is an unannealed Ti-Au. The p-type contacts have apertures around the photonic crystals that allow the photoluminescence to be collected. The apertures are shown in the SEM image of a typical device in Fig. 7.9b.

7.3.2 Device Physics

The p-type and n-type material in the P-I-N diode creates an electric potential across the intrinsic layer. In our structures this intrinsic layer contains the InAs QDs.

By altering the electric field across the QD the bandgap and potential well are modified. Single particle energy levels within the QD are lowered . This is known as the *quantum confined Stark effect*.

The resultant field across the QD is a sum of the intrinsic field produced by the P-I-N diode structure, V_0, and the externally applied fields.

$$V_{Resultant} = V_0 + V_{applied} \tag{7.2}$$

In our structures the electric field parallel to the direction of the QD growth can be controlled by both the magnitude of an externally applied voltage and the screening of the intrinsic field by carriers generated through above bandgap optical excitation.

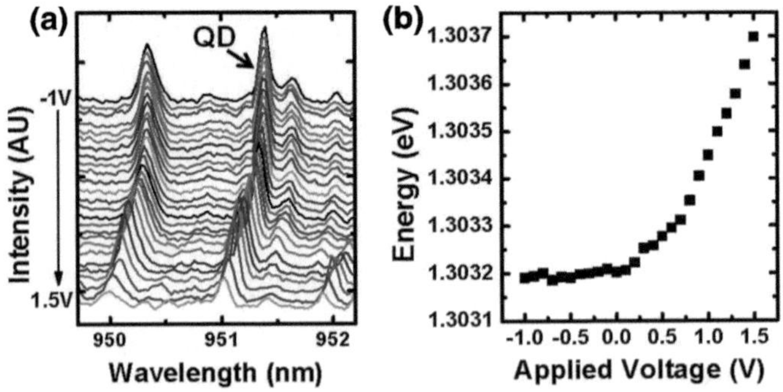

Fig. 7.10 **a** QD PL emission as a function of applied voltage. **b** Energy of the PL emission labelled QD in figure (**a**) as the externally applied voltage across a P-I-N diode is varied between −1 and 1.5 V. A maximum Stark shift of 0.5 meV is observed

7.3.3 Electrically Controlled Tuning

A tunable diode laser resonant with a higher order cavity mode excites the QDs within the photonic crystals. Subsequent PL was collected and recorded by a CCD after being dispersed by a spectrometer.

As the applied bias increases from $V = 0$ to 1.5 V the internal field across the QD is reduced. In Fig. 7.10 a Stark shift of 0.5 meV is observed for a typical QD emission line for an external applied voltage between −1 and 1.5 V. The range of applied field is limited by the electrical turn on of the diode under forward bias at around 1 V and quenching of PL emission due to tunnelling of carriers out of the QD at a reverse bias of −1 V.

7.3.4 Optically Controlled Tuning

By exciting the QDs within the photonic crystal diode structure using a HeNe laser emitting at 633 nm (1.96 eV) (in addition to the resonant excitation occurring through the higher order mode) carriers are generated above the bandgap of the intrinsic GaAs layer (1.43 eV) so that they are no longer localised within the quantum dot. The delocalised carriers subsequently act to screen the internal field across the QD.

By controlling the number of above bandgap photogenerated carriers created through the power of incident HeNe laser light, an optically controlled Stark Shift of 0.8 meV was observed as shown in Fig. 7.11.

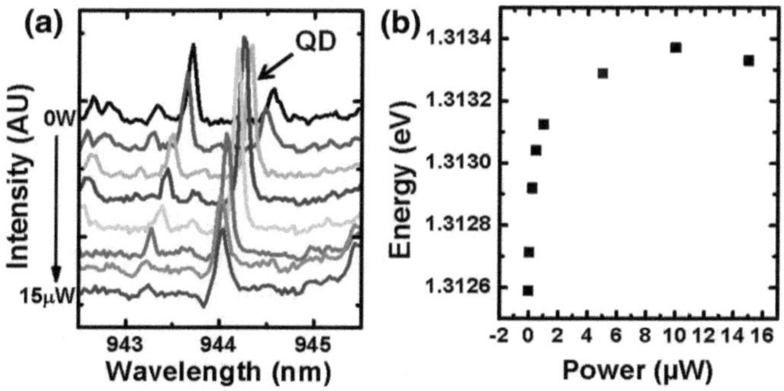

Fig. 7.11 **a** PL emission from a photonic crystal P-I-N diode structure under resonant and above bandgap excitation. As the power of the above bandgap excitation increases charged photogenerated carriers screen the internal field and induce a Stark Shift. **b** Energy of quantum dot emission labelled QD in figure (**a**) as a function of the incident HeNe laser power a Stark Shift of 0.8 meV is observed

7.3.5 Discussion

The observed Stark shifts of 0.5 and 0.8 meV for the electrically and optically controlled tuning represent a useful magnitude but are somewhat short of the best results achieved elsewhere [12].

The reason for this may be due to the density of the p-type doping. Our p-type material was Be doped to a density of 2×10^{18} cm^{-3} whilst elsewhere the p-type material that was C doped to 2×10^{19} cm^{-3}.

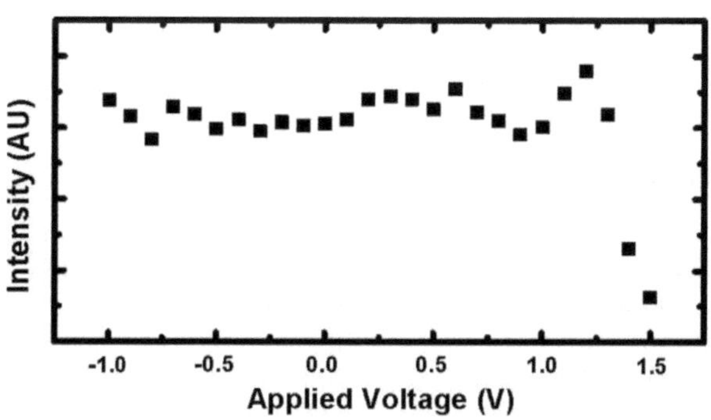

Fig. 7.12 Intensity of the emission line labelled QD in Fig. 7.10a as the applied voltage is varied between −1 and 1.5 V. There is little change in the emission intensity of around 2,000 counts for voltages below 1 V before the electrical turn on of the diode is reached at which point the intensity rapidly decreases

Two experimental results also led to concerns that the lower doping in our p-type layer was resulting in the poor performance, by facilitating only a small external field across the QDs within the cavity.

When the P-I-N diode is under high reverse electric field we expect to see a drop in PL intensity due to the increase in recombination lifetime. However, Fig. 7.12 shows that for a QD emission line in our sample we see little change in the intensity of emission before we begin to draw current and observe heating effects. This suggests we are drawing current and applying field only through the part of the semiconductor material that lies directly below the contacts on the surface of the diode. Only a small electric field is being induced across the QDs in the cavity which would account for the small Stark shift.

The diode characteristics at low temperatures confirms there is a large 'freeze out' of carriers as the temperature is reduced. When the temperature reduces to a value equivalent to the energy required to promote carriers from the acceptor levels of the p-type layer into the conduction band it results in an increase of contact resistance and degradation of the diode characteristics at low temperatures. In an attempt to improve the observed Stark shift two attempts were made to improve the performance of the p-type layer.

A semi-transparent p-type contact was evaporated over the whole diode surface covering the top of the photonic crystals in an effort to improve the external electrical field induced across the QDs in the cavity. This was unsuccessful as no PL from the QDs in the cavities was subsequently observable after the evaporation.

For a second attempt, a second wafer was grown with a higher doping density in the p-type region, 2×10^{19} cm^{-3}. This approach proved promising as the diode characteristics were much improved at lower temperatures suggesting that less hole 'freeze out' was occurring. However, we were unable to observe any PL emission from QDs in the wafer and expect the cause of this to be an error in the wafer growth process.

7.3.6 Further Development

A new design of electrically tuned QD emission has been developed. QDs have been fabricated in potential wells that are made from material with a larger bandgap than the QD material [13]. This has increased the magnitude of the electric field that can be applied before carrier tunnelling quenches PL emission. This additional carrier confinement achieves a Stark shift of 25 meV, an increase of ten times on the P-I-N structure.

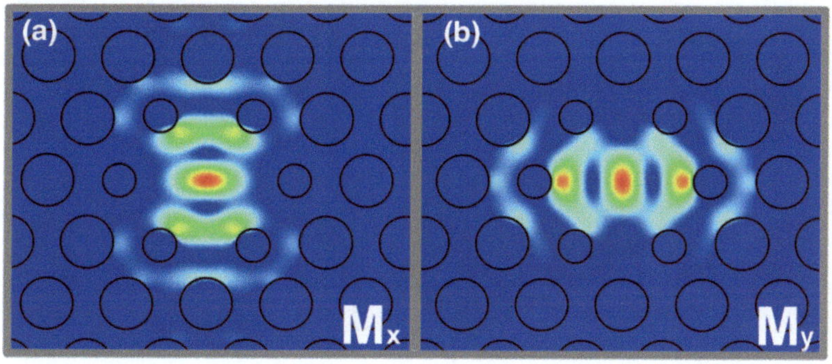

Fig. 7.13 Calculated electric field distributions for the **a** x-polarised and **b** y-polarised fundamental modes of the H1 PC cavity

7.4 Restoring Mode Degeneracy in H1 Photonic Crystal Cavities by Uniaxial Strain Tuning

We have shown how electrical tuning of QD emission energy can be achieved in a P-I-N diode and have also discussed that cavity mode tuning is also an important technology for the development of quantum optical systems. Here an in-situ technique for controlling the mode splitting in H1 PC cavities using an externally applied strain is presented.

Many photonic crystal cavity designs involve some type of line defect [14, 15], which results in linear polarisation of the cavity modes. This has a strong effect on the polarisation of photons emitted by QDs coupled to the cavity modes [16]. In certain applications, it is essential that the cavity does not influence the polarisation of the emitted photons. For example, to determine the spin state of a single electron in a QD, it is necessary to detect a circularly polarised photon. This can be achieved by using a pair of degenerate modes with orthogonal polarisations, as expected from H1 [17] and Calzone [9] single hole defect cavities and also with coupled nanobeam cavities [18].

Figure 7.13 shows the electric field intensities at the centre of the slab for two fundamental modes of an H1 cavity, M_x and M_y, calculated by finite difference time domain (FDTD) simulations.

Despite being theoretically degenerate, the degeneracy of the modes is broken by imperfections in the fabrication. This results in the observation of two linearly polarised fundamental modes with a typical mode splitting, $\Delta\lambda = 1 - 2\,\text{nm}$ [9, 17] This mode splitting can be corrected with a post-fabrication tuning technique [9] or by compensating for the fabrication errors in the cavity design [19].

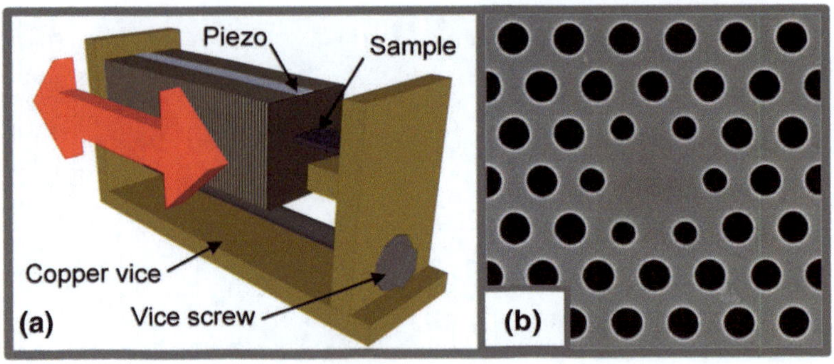

Fig. 7.14 **a** Schematic diagram of the experimental set-up used to apply a uniaxial strain to the sample. **b** Scanning electron microscope image of an H1 PC cavity with the optimised parameters for hole size and position employed for the experimental studies

7.4.1 Device

We have developed an in-situ technique for reversible fine tuning of the mode splitting in an H1 PC cavity from ~0.5 nm to less than 0.15 nm by using an externally applied uniaxial strain [20]. The application of a uniaxial strain deforms the photonic lattice sufficiently to compensate for fabrication errors and restore mode degeneracy. Figure 7.14a shows a schematic diagram of the experimental apparatus. A copper vice is mounted on the cold finger of a liquid helium flow cryostat. The sample is mounted at one end of the vice, with an 18 mm long piezoelectric actuator at the other. The vice screw is used to tightly secure the sample and actuator at room temperature, before the whole arrangement is cooled to ~15 K for the μPL experiments.

As the sample cools, the copper thermally contracts increasing the strain on the sample. During the measurements, the strain can then be further increased by increasing the bias voltage applied to the piezoelectric actuator, V_p, or reduced with a negative voltage.

7.4.2 Sample

The wafers used in this study consist of a 130 nm GaAs membrane with a single layer of nominally InAs QDs at its centre. Figure 7.14b shows a SEM of a typical H1 cavity used in this work. The photonic crystal had lattice period, $a = 260$ nm and hole radius, $r = 0.29a$. The six holes nearest to the missing hole defect were shifted out by a distance $s_i = 0.1a$ and had a reduced radius, $r_i = 0.7r$.

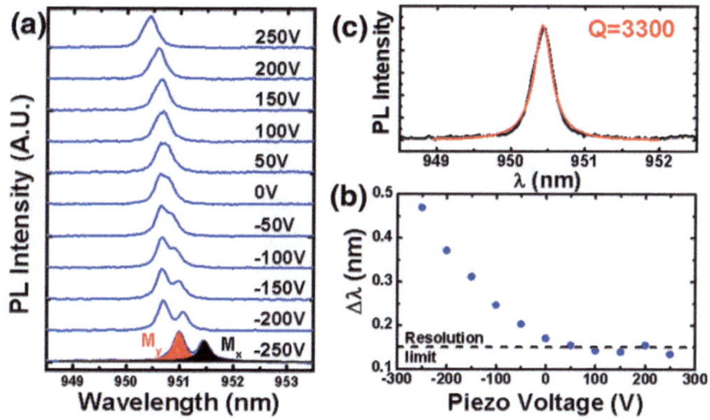

Fig. 7.15 **a** μPL spectra recorded from an H1 PC cavity as a function of piezo voltage from -250 to 250 V. **b** Cavity mode splitting as function of piezo voltage. **c** Unpolarised fundamental mode emission at a bias of 250 V, with a Q-factor of 3,300

7.4.3 Strain Dependent PL

Figure 7.15a shows the strain dependent PL spectra of the fundamental modes of the H1 cavity. With $V_p = -250\,\text{V}$, the mode splitting is initially $\sim$0.47 nm, with M_x at higher wavelength than M_y. As the strain in the sample changes M_x is tuned more strongly than M_y, showing a $\sim$3 times larger change in wavelength than M_y and a total tuning range of $\sim$1 nm.

This is expected for a uniaxial strain in which the photonic lattice is deformed more in one direction than the other and is similar to previous measurements where the photonic lattice is stretched in one direction [19].

This difference in the tuning rate of the two modes results in a change in $\Delta\lambda$, as plotted in Fig. 7.15b. As V_p is increased, $\Delta\lambda$ is reduced from $\sim$0.47 to < 0.15nm for $V_p > 200\,\text{V}$. It is not possible to resolve a splitting below this value as the linewidth of the cavity modes is $\sim$0.25 nm. The splitting shows a non-linear dependence on the voltage applied to the piezo-actuator, probably related to hysteresis in the displacement, as expected for this type of piezo-actuator. Figure 7.15c shows the spectrum at $V_p = 250\,\text{V}$, with degeneracy of the modes restored and a Q-factor of 3,300.

7.4.4 Further Work

The results demonstrate that the application of strain is an effective method for tuning the mode resonances of a PC cavity. Nevertheless, a drawback with this technique is that it requires the uniaxial strain to be applied to the entire sample. It would be more effective to apply the strain only to the PC. This could be achieved by using a micro-electro-mechanical approach. For example, an in-plane electro-static force could be used to apply strain to the PC membrane. This is likely to be more effective in the case of crossed nanobeam cavities [18], as the force required to deform a nanobeam will be less than for a 2D PC. The ability to control mode splitting with very high accuracy has the potential to be used in conjunction with QDs for entangled photon pair generation. It has been proposed that a PC cavity can be used to overcome the fine structure splitting (FSS) inherent in self-assembled QDs [21]. This proposal relies on matching the mode splitting of the cavity to the FSS of the QD in the strong-coupling regime. Strain tuning can provide the control necessary to achieve this.

References

1. K. Hennessy, A. Badolato, M. Winger, D. Gerace, M. Atatüre, S. Gulde, S.Fält, E.L. Hu, A. Imamoğlu, Quantum nature of a strongly coupled single quantum dot-cavity system. Nature **445**(7130), 896–899 (2007)
2. K.H. Lee, A.M. Green, R.A. Taylor, D.N. Sharp, J. Scrimgeour, O.M. Roche, J.H. Na, A.F. Jarjour, A.J. Turberfield, F.S.F. Brossard, D.A. Williams, G.A.D. Briggs, Registration of single quantum dots using cryogenic laser photolithography. Appl. Phys. Lett. **88**(19), 193106 (2006)
3. A. Dousse, L. Lanco, J. Suffczyński, E. Semenova, A. Miard, A. Lemaître, I. Sagnes, C. Roblin, J. Bloch, P. Senellart, Controlled light-matter coupling for a single quantum dot embedded in a pillar microcavity using far-field optical lithography. Phys. Rev. Lett. **101**(26), 30–33 (2008)
4. A. Dousse, J. Suffczyński, R. Braive, A. Miard, A. Lemaître, I. Sagnes, L. Lanco, J. Bloch, P. Voisin, P. Senellart, Scalable implementation of strongly coupled cavity-quantum dot devices. Appl. Phys. Lett. **94**(12), 121102 (2009)
5. S.M. Thon, M.T. Rakher, H. Kim, J. Gudat, W.T.M. Irvine, P.M. Petroff, D. Bouwmeester, Strong coupling through optical positioning of a quantum dot in a photonic crystal cavity. Appl. Phys. Lett. **94**(11), 111115 (2009)
6. A. Laucht, J. Villas-Bôas, S. Stobbe, N. Hauke, F. Hofbauer, Mutual coupling of two semiconductor quantum dots via an optical nanocavity mode. Arxiv preprint arXiv:0912.3685: 1–12 (2009)
7. K. Hennessy, A. Badolato, A. Tamboli, P. M. Petroff, E. Hu, M. Atatüre, J. Dreiser, A. Imamoğlu, Tuning photonic crystal nanocavity modes by wet chemical digital etching. Appl. Phys. Lett. **87**(2), 021108 (2005)
8. S. Strauf, M.T. Rakher, I. Carmeli, K. Hennessy, C. Meier, A. Badolato, M.J.A. DeDood, P.M. Petroff, E.L. Hu, E.G. Gwinn, D. Bouwmeester, Frequency control of photonic crystal membrane resonators by monolayer deposition. Appl. Phys. Lett. **88**(4), 043116 (2006)
9. K. Hennessy, C. HoÌĹ gerle, E. Hu, A. Badolato, A. ImamogĹĘ lu, Tuning photonic nanocavities by atomic force microscope nano-oxidation. Appl. Phys. Lett. **89**(4), 041118 (2006)
10. A. Faraon, D. Englund, I. Fushman, J. Vučković, N. Stoltz, P. Petroff, Local quantum dot tuning on photonic crystal chips. Appl. Phys. Lett. **90**(21), 213110 (2007)
11. A. Faraon, A. Majumdar, H. Kim, P. Petroff, J. Vuckovic, Fast electrical control of a quantum dot strongly coupled to a nano-resonator. Opt. Express **17**(21), 18651–18658 (2009)
12. A. Laucht, F. Hofbauer, N. Hauke, J. Angele, S. Stobbe, M. Kaniber, G. Böhm, P. Lodahl, M.-C. Amann, J.J. Finley, Electrical control of spontaneous emission and strong coupling for a single quantum dot. New J. Phys. **11**(2), 023034 (2009)
13. R. Patel, A. Bennett, I. Farrer, C. Nicoll, D. Ritchie, A. Shields, Tunable indistinguishable photons from remote quantum dots. Arxiv preprint arXiv:0911.3997: 1–11 (2009)
14. B.-S. Song, S. Noda, T. Asano, Y. Akahane, Ultra-high-Q photonic double-heterostructure nanocavity. Nat. Mater. **4**, 207–210 (2005)

15. Y. Akahane, T. Asano, B. Song, S. Noda, High-Q photonic nanocavity in a two-dimensional photonic crystal. Nature **425**, 944–947 (2003)
16. R. Oulton, B.D. Jones, S. Lam, D. Szymanski, D. Sanvitto, A.M. Fox, D.M. Whittaker, M. Hopkinson, M.S. Skolnick, Polarized quantum dot emission from photonic crystal nanocavities studied under mode- resonant enhanced excitation. Opt. Express **15**(25), 17221–17230 (2007)
17. M. Shirane, S. Kono, J. Ushida, S. Ohkouchi, N. Ikeda, Y. Sugimoto, A. Tomita, Mode identification of high-quality-factor single-defect nanocavities in quantum dot-embedded photonic crystals. J. Appl. Phys. **101**(7), 073107 (2007)
18. K. Rivoire, S. Buckley, J. Vučković, Multiply resonant high quality photonic crystal nanocavities. Appl. Phys. Lett. **99**(1), 013114 (2011)
19. I.J. Luxmoore, E.D. Ahmadi, A.M. Fox, M. Hugues, M.S. Skolnick, Unpolarized H1 photonic crystal nanocavities fabricated by stretched lattice design. Appl. Phys. Lett. **98**(4), 041101 (2011)
20. I. J. Luxmoore, E. D. Ahmadi, B. J. Luxmoore, N. A. Wasley, A. I. Tartakovskii, Restoring mode degeneracy in H1 photonic crystal cavities by uniaxial strain tuning, Appl. Phys. Lett. **121116**, 1–4 (2012)
21. R. Johne, N. Gippius, G. Pavlovic, D. Solnyshkov, I. Shelykh, G. Malpuech, Entangled photon pairs produced by a quantum dot strongly coupled to a microcavity. Phys. Rev. Lett. **100**, 1–4 (2008)

Chapter 8
Conclusions and Future Directions

8.1 Summary

This thesis has described the spectroscopic measurements used to investigate how photonic crystal waveguides and InP QDs could be applied within integrated quantum optical circuits and their current limitations. It has also presented a crossed waveguide device capable of acting as a QD spin photon interface in a planar geometry and discussed a number of technical developments that have or will allow us to fabricate increasingly complex integrated quantum optical circuits with added functionality.

Chapter 1 presented an introduction to quantum information processing and the proposed implementation in the solid state using III-V semiconductors. The physics of photonic crystals, quantum dots and cavity QED on which the work presented in the subsequent experimental chapters is based was also introduced.

Chapter 2 described the techniques that were employed during the experiments presented in the thesis. The details of sample fabrication, as well as the set-ups used for standard optical spectroscopy measurements were presented including a brief description of the two cryostat systems that were used. Details of the experimental method used to provide two spatially selective collection paths, interference and two-colour detection was also presented.

A summary of experimental results and the conclusions drawn from the subsequent chapters is presented here:

Chapter 3: Photonic Crystal Waveguides

By studying the Fabry-Perot resonances present in photonic crystal waveguides of three lengths, the photon propagation lengths were deduced as a function of the group index for the bound waveguide mode. It was found that the photon propagation length in GaAs QD PCWs are currently limited to a few tens of microns by unintentional disorder when operating in the slow-light regime. The presence of disorder within the photonic crystal was supported by observations of Anderson localised modes at

N. A. Wasley, *Nano-photonics in III–V Semiconductors for Integrated Quantum Optical Circuits*, Springer Theses, DOI: 10.1007/978-3-319-01514-9_8,

high group indices. This has significant ramifications for the use of slow-light based PCWs in integrated quantum optics where photon propagation lengths many times greater may be required.

However, a number of possible applications for slow-light based PCWs in hybrid quantum optical circuits were presented including PCW phase shifters and engineered PCWs for flat-band slow-light.

Chapter 4: On-chip Interface for In-plane Polarisation Transfer

A scheme for interfacing an optically addressed QD spin qubit to a path encoded photon using a crossed waveguide device was presented. The polarisation state of a circularly polarised photon was recovered at a remote location using an in plane interferometric measurement. This opens up the possibility of a highly scalable quantum optical network compatible with conventional semiconductor processing techniques. This work provides a promising route toward constructing a scalable on-chip network of spins connected by optical waveguides.

Chapter 5: Direct In-plane Readout of QD Spin

We observed that the crossed waveguide device operates in two regimes dependent on the location of the QD and have shown that the device can be used for direct read-out of spin information as well as in-plane transfer.

The observations were accompanied by FDTD simulations that were able to simulate the experimental observations and a more intuitive description of the crossed waveguide structure as a combination of virtual beamsplitters was also provided.

Chapter 6: Fourier Transform Spectroscopy of InP QDs in Photonic Crystal Cavities

By means of Fourier transform spectroscopy we were able to analyse the dephasing processes within the InP QDs. Whilst photonic crystal modes may be a possible route to reducing the contribution to dephasing caused by acoustic phonons, we have found that fluctuations due to trapping of carriers still limits the coherence time for InP QDs.

Chapter 7: Development of Additional Technological Approaches

This chapter presented three technical developments that have important consequences for the development of on-chip quantum optical circuits. (i) A quantum dot registration technique that identifies the position of single quantum dot to <50 nm accuracy through optical spectroscopy. (ii) The tuning of the QD transition energy by P-I-N diodes via the quantum confined Stark effect. (iii) Restoration of the degeneracy of the two fundamental modes of a H1 type PCC by a strain tuning technique using a piezo actuator.

8.2 Summary of Further Work

A number of possible directions for future work or details of the work required to further an investigation have been suggested as part of each chapter. Where appropriate the details of further experimental work that has been undertaken or interesting preliminary results that lead on to the possibility of future research, have been also detailed as part of the experimental chapters.

I will summarise here the most interesting avenues for future work and the work currently being undertaken in the group that has already been discussed. I will then present a more detailed discussion of a direction in which long term development may take place as a result of the experiments performed in this thesis, not yet formally discussed.

8.2.1 Chapter 3: Photonic Crystal Waveguides

The guided mode of a PCW is sensitive to hole size and crystal geometry. By modifying the position and shape of the air holes the group index experienced by the propagating photon may be tuned.

This observation leads directly to a number of applications. The first is as a solution to high losses in the slow-light mode, whereby a hybrid photonic crystal circuit is employed in which the periodicity or r/a value of the crystal is varied such that the propagating mode switches between slow-light and fast-light propagation at a fixed wavelength. Other applications include:

PCW On-chip Phase Shifter

There are a broad range of applications for quantum optical circuits that require active control of the photon path. For example on-chip interferometry, where the ability to control and set the phase delay of a photon path is required. This active control of slow-light on-chip also has applications in optical buffering and dynamic dispersion compensation.

By temperature tuning the PCW, active control of the group index may be achieved.

To demonstrate this principle, Fabry-Perot resonances were recorded in a PCW as a function of temperature. The temperature of the sample was externally controlled by a heating element in the cold finger cryostat. Analysis of the recorded spectra for each temperature allowed the group index of the waveguide mode to be mapped and evidence of a controllable phase shift to be observed.

Future work in developing this application will use the observed control over the group index to demonstrate control over the phase of propagating photons. This will likely require an interference measurement. The stability of the temperature control will also require more detailed analysis and investigation to determine the resolution of the temperature control.

PCW Engineering

In addition to the active control of PCW elements, the sensitivity of the waveguide mode to the PCW geometry provides an opportunity to permanently engineer desired characteristics into the waveguide mode by controlling the design of the photonic crystal.

One aim of modifying the W1 PCW design is to reduce dispersion and provide a constant interaction strength between the QDs and the waveguide mode over a larger wavelength window than currently exists in PCWs or typically available in PCCs.

Modifications to the hole positions of the first two lattice periods either side of the waveguide were made by R. J. Coles and a flat-band region of moderate group index of around 30 was observed between 935 and 945 nm.

Calculations of the Purcell factor conducted by R. J. Coles for these flat band devices suggest that a Purcell factor (PF) = 4 could be achieved in a 6 nm bandwidth and even a PF = 10 for 1 nm bandwidth, although a large group index of ~70 would be required.

Work has also begun on using longitudinal shifts of the first row of holes as this can increase the Purcell factor for a given group index. With the possibility of achieving PF = 10 at $n_g = 50$, which would be a considerable improvement on the transverse shifted PCWs.

8.2.2 Chapter 5: Direct In-plane Readout of QD Spin

Spin-memory and Photon Emission Path Control

By resonantly pumping the QD via the sharp resonance that occurs at a detuning of one longitudinal optical (LO) phonon above the QD state it is possible to pump a single transition of the Zeeman split doublet, based on the polarisation of the excitation laser.

By employing this excitation technique we made a preliminary demonstration of a system in which the spin preparation technique pre-determines the emission path of the photon emitted from the QD.

Although the recorded data demonstrated a good control over the emitted photon path, the recorded contrast suggested that it is the preparation of the spin state that is limiting the control over the deterministic excitation to which path relationship.

Improving this spin memory technique will require a resonant pumping technique, which will improve the QD preparation fidelity. This technique will require both a tunable laser and improved spatial filtering in the collection path. A tunable laser capable of providing the resonant excitation required for further investigation of this system is now available in the lab.

8.2.3 Chapter 7: Technical System Developments

QD Registration of Two Spectrally Resonant QDs

Although a successful registration and writing of a photonic pattern around the registered QD has not as yet been completed, there are already a number of exciting applications such a capability could provide. In wafers with a relatively high QD density there is a reasonable probability that when mapping the marked area, QD emission from more than one QD may be recorded in resonance with each other, such that the emission peaks are less than the linewidth apart. This creates the possibility of developing a system that deterministically couples quantum dots via photonic crystals or plasmonic waveguides.

8.3 Future Direction: An Integrated Circuit to Transfer an Arbitrary QD Spin State to a Photon Superposition State

The development of on-chip integrated quantum optical circuits from III-V materials is a goal not just for our research group but for the research fields of nano-optics, photonics and quantum information processing. The future direction for the work that has been presented in this thesis is to continue toward the integration of optical components and increasing the available functionality through on-chip demonstrators. As such, I will describe a direction of work that I feel builds on the components presented in this thesis whilst maintaining progress toward the demonstration of an integrated optical circuit for QIP. This direction offers a number of intermediate stages of development that represent important progress in their own right.

As discussed, quantum dot systems containing electron spins may play a role in a future photonic quantum optical circuits as a means of storing and manipulating a quantum state in a spin superposition, to be transferred to a photon polarisation superposition.

Our collaborators at The University of Bristol, R. Oulton and A. Young have demonstrated that any arbitrary superposition state of dipoles maps to a well-defined path-encoded photon state in the crossed nano-wire device presented in Chap. 4. This implies that when photons are emitted from a charged QD, leaving behind a spin state, it gives rise to photon-spin entanglement.

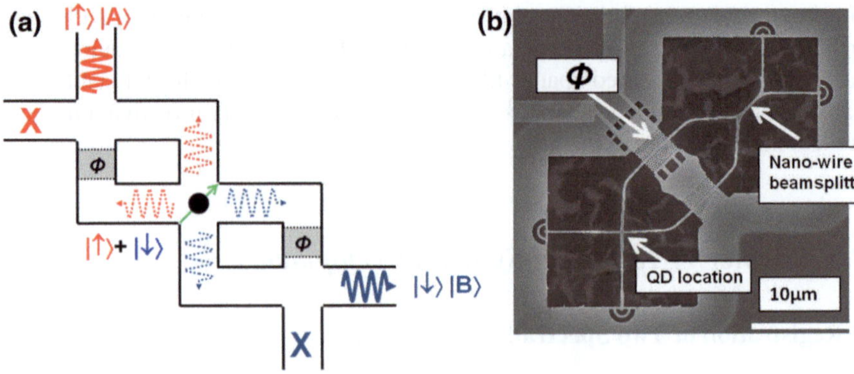

Fig. 8.1 A schematic diagram of an on-chip spin-photon entangler for photons to be used as inputs into a linear optical circuit. In the case of where the QD is located as in Device-B, the O_T, O_L and O_B, O_R waveguide arms may be combined using a standard Mach-Zehnder type device, where the Φ *box* represents a tunable phase shift. The output may be tuned such that the photon component is emitted out of one arm only. What results is two photon paths that may be fed into a typical linear-optical circuit. (**b**) SEM image of a prototype quantum optical circuit that shares similar geometry to the schematic in (**a**) with the incorporation of an on-chip PCW phase shifter and beamsplitter into the O_B, O_R output arms of a crossed waveguide device

In particular, for the Device-B, they have shown that the O_T, O_L and O_B, O_R waveguide outputs are identical for any dipole superposition state. This is particularly advantageous, as the O_T, O_L and O_B, O_R outputs may be combined to give just two paths representing the superposition state.

The fact that a charged QD can emit a photon whose path encoding has a one-to-one correspondence does not only mean that it is possible to read out the spin on-chip, but it may also be used as a spin-photon interface, where the spin is entangled with a photon. This is the starting point for many possible on-chip quantum photonic devices.

Figure 8.1 shows a schematic of a possible device that could be used in scalable linear-optical circuits. The photon output into each arm are entangled with the spin in the QD. By feeding the outputs of many sets of these modules into a linear-optical circuit it would allow interaction of photons from several QDs. This could be used to provide remote spin entanglement by interfering sets of output states.

The SEM image displayed in Fig. 8.1b was presented in Chap. 3. It is a prototype device that incorporates a number of the elements required to realise the on-chip spin-photon entangler shown in Fig. 8.1b. Although it only consists of half of the fully functional module, this intermediate circuit offers flexibility within the development process, whilst maintaining the consistency of studying a single structure design.

The circuit is largely fabricated from elements featured in this thesis and designed from the conclusions drawn from our measurements. The structure is based around the crossed waveguide structures presented in Chaps. 4 and 5. It consist of a hybrid nano-wire / PCW system which is a result of the investigations in Chap. 3, and the tunable phase shift is similar to the PCW type device presented in Chap. 3.

An important element in this structure is the nano-wire beam-splitter, which is a 50:50 coupler with beam-splitter properties. It has been widely used elsewhere, however, it is currently the least understood and little tested element in our circuit. Work is currently being undertaken in our group to devise and optimise a 50:50 beam-splitter suitable for use in this type of circuit, which is required prior to any further development of this system.

The purpose of this structure is to test functionality and aid the integration of the elements required to fabricate a on-chip spin-photon entangler. With four out-couplers on the structure, it allows the structure to be studied with increasing experimental complexity. This flexibility is why the structure offers a promising route toward the development of integrated circuits.

At its simplest, experimental measurements can be conducted by exciting the device through the out-couplers on the left hand side of the SEM image and collecting the excitation laser from the out-couplers to the right of the image once it has propagated through the circuit. This experiment will isolate the beam-splitter and allow it to be studied. The same excitation and collection scheme can also be used to investigate the performance of the phase shifter.

Once satisfied by the performance of the beam-splitter and phase shifter, another level in complexity may be added by incorporating the QD spin-photon interface by exciting the QD located at the centre of the structure. The spin-photon interface as studied in Chaps. 4 and 5 coupled with the control over the photon path given by the phase shifter and beam-splitter begins to have a similar structure to part of the on-chip spin-photon entangler displayed in Fig. 8.1a.

It may also be possible to resonantly excite the QD via the out-couplers or introduce two more spatial collection paths and perform the path controlling experiment simultaneously off-chip, and subsequently, once the module has been demonstrated, QD registration and charge tuning techniques could be used to add further functionality.

The flexible functionality provided by this structure is an important advantage when considering how the development of an on-chip spin-photon entangler module and other integrated circuits is to proceed.

About the Author

Nicholas A. Wasley

1997–2005:	St. Ninian's High School, Isle of Man
2005–2009:	The University of Manchester MPhys Physics—First Class Honours
2009–2013:	The University of Sheffield PhD: Nano-photonics in III-V semiconductors for integrated quantum optical circuits
2013–Present:	Sagentia, Consultant Physicist

As an Isle of Man Government awarded Russell-Willis Scholar, Nicholas obtained an MPhys Physics degree with first class honours from the University of Manchester in 2009.

He subsequently conducted research for his PhD as a member of the Low Dimensional Structures and Devices group in the Department of Physics and Astronomy at the University of Sheffield under the supervision of Prof. M. Skolnick (FRS), completing his thesis in January 2013.

Nicholas has co-authored five peer reviewed papers and presented work at a number of conferences including ICPS 2012 where he was nominated for the Young Scientist Best Paper Award.

Nicholas currently works as a consultant physicist for Sagentia, a global innovation, technology and product development company that provides out-sourced R&D consultancy services in the medical, industrial and consumer sectors based in Harston, Cambridge.

N. A. Wasley, *Nano-photonics in III–V Semiconductors for Integrated Quantum Optical Circuits*, Springer Theses, DOI: 10.1007/978-3-319-01514-9,

MIX
Papier aus verantwortungsvollen Quellen
Paper from responsible sources
FSC® C105338

If you have any concerns about our products,
you can contact us on
ProductSafety@springernature.com

In case Publisher is established outside the EU,
the EU authorized representative is:
Springer Nature Customer Service Center GmbH
Europaplatz 3, 69115 Heidelberg, Germany

Printed by Libri Plureos GmbH
in Hamburg, Germany